Math for Inquiring Minds:
A Collection of Authentic Vignettes, Curious Questions and Classic Problems for Secondary Students

Susan Culver

Copyright 2017

Introduction

Welcome to **<u>Math for Inquiring Minds: A Collection of Authentic Vignettes, Curious Questions and Classic Problems for Secondary Students</u>**! Following is the reason I wrote this problem-solving workbook: 1. "A family goes into a store and buys 6 pair of socks and 2 pairs of mittens for $26. They decide the next day that they need more of these items, so they buy 4 pairs of socks and 7 pairs of mittens for $40. How much does a pair of socks and a pair of mittens cost?" READ THE PRICE TAG. Or 2. "Jane is 2 times older than Bill. Four years ago, she was 11 years older than Bill. How old is Jane?" ASK HER.

I believe these two example questions are one of the reasons that students may find math frustrating. Both of these problems seem "real-world," but they are not. No one realistically would use algebra to solve these, yet most math textbooks and many math teachers will present these problems in this way in a classroom. Each of the problems in this book is designed to stamp out useless or boring math story problems that pose as real-world problems, but really are not. Each of these problems must have at least one of the following qualifications: 1. The problem could very easily occur in the real world naturally or be set up easily. 2. It is a math question a student may have pondered on their own. 3. A student would find the solution to the problem makes sense, is useful, is interesting, or is counterintuitive (and so, interesting). 4. The student may need to find additional real-world data to solve the problem. 5. The problem naturally invites discussion while solving and when the solution is found. 6. The problem may help tell the story of the current conditions and wonders of our world. 7. The problem may be designed to open eyes and change behavior. 8. The problem is fun!

Though I am nearly certain that as an educator, you or your school/district follow a specific method for problem-solving, I also offer my own thoughts.

1. When possible, use these problems in an environment that invites collaboration, discussion, perseverance and presentation. 2. Emphasize visualization – the problems are a natural fit for this strategy. Use real student names in each problem to help. 3. Focus on associating all values with units, using units to help set up the problem properly, and including units in the solutions. 4. Have discussions on what missing data may be required to solve the problem. If a problem requires outside data, you may provide it or ask students to locate the information. *If no outside data is needed, then the internet should not be used as a resource.* And, of course, 5. Does the answer make sense, or if it seems counterintuitive, why?

      Each provided solution includes the following information in this order: A. Outside data critical to the problem, with an example provided that is relevant to the author. B. Math concepts or strategies. You can use these key words to search for certain types of problems to offer students on warmups, assignments or assessments. Typical key words include: rates, conversion, proportions, formulas, probability, trigonometry, linear systems, quadratics, percents, exponential, financial, distance, etc. C. Solutions or example solutions using provided example data.
D. Work. I record intermediate values to the 1/1000s place and final values to the 1/10s place or to the logical place.

      Please, as one math educator/enthusiast to another, as the purchaser of this document, please feel free to produce worksheets/warm-ups/assessments for your students. Please, however, do not copy this book or documents you create from it to share with educators/enthusiasts who have not purchased it. The low price is meant to allow wide-spread enjoyment and strengthen problem-solving skills for students of <u>each</u> educator who buys it.

It's an honor to me that you have chosen this collection as a math resource!

If you would like a free electronic file of this book to produce your own documents from these problems, please contact me at mathceptional@yahoo.com with proof of your purchase of this book from Amazon.

Math for Inquiring Minds: A Collection of Authentic Vignettes, Curious Questions and Classic Problems for Secondary Students

1. **Sailing away – a little longer:** You are navigating your sailboat from St. Joseph, Michigan, to Milwaukee, Wisconsin. Thirty-four nautical miles into your trip, a small storm has popped up in the middle of your direct path, and so you must navigate 32 degrees off the direct course for 20 nautical miles to sail around this storm. How many nautical miles longer will your trip be than your direct path?

2. **Waiting in line:** You are waiting in line for an amusement park ride. You are the 20th person in line, and your friend, who stopped to get a snack, is 100th in line. How many people are between you both?

3. **How old is your Christmas tree?:** It is your family's tradition the weekend after Thanksgiving each year to visit a Christmas tree farm and bring a tree home to decorate. This year, you choose a blue spruce that is 7 feet 5 inches tall. Approximately what is the age of your tree in years?

4. **That is one tall tree:** You are standing in the sunshine in your aunt's backyard which has a huge cottonwood tree. Your shadow is 3 feet long. You measure and find the shadow of the cottonwood tree is 49 feet long, how tall is the tree?

5. **Late for school?:** Your home is 1.2 miles from school. As you are walking to school, you realize at the halfway point, you forgot an important assignment that is due on your kitchen table at home. You decide to run home to get the assignment and run to school. It is 7:35 am and

school starts at 7:45 am. If you can run at 12 mph, will you make it to school on time?

6. Storm watch: You are at your friend's house swimming in the pool when you see a storm cloud approaching and a flash of lightening. Five seconds later you hear thunder. If the storm is approaching at 25 miles per hour, when will the storm be at your friend's house?

7. Free throw percentage: Your basketball coach encourages each player to have a free throw percentage of 50% or higher. During free throw practice, you currently have made 5 of 16 shots. How many consecutive free throws do you have to make in a row so you are at 50 percent?

8. High school musical: You are working on the scenery for your high school's musical for this year. You are working on your final project -- a forest scene. You have 10 more hours of auditorium time available to you before the first showing, but you realize this scene will take you at least 18 hours to finish. You call your little sister for assistance, who would be able to finish this project on her own in 24 hours. Working together, will you both finish the project on time?

9. Wheels on the bus: You are on a bus on a long fieldtrip to The Henry Ford when you are mesmerized by the turning wheels on the car next to you. If the car is going just as fast as the bus -- 60 mph -- and the wheels on the car are a standard 2 feet in diameter, how many revolutions do the car's tires rotate per minute?

10. Ribbon cutting ceremony: You are earning service hours by making 22 ribbon necklaces for Mu Alpha Theta medals for recipients to wear during an upcoming ceremony. If each ribbon needs to be 18 inches long, and you have 4 rolls of ribbon that are 6-2/3 feet long, do you have enough ribbon to finish the project?

11. **Ladder safety:** According to safety standards, the safest position for climbing a ladder leaning against a wall is when the horizontal distance between the top of the ladder and the foot of the ladder are 1/4 the length of the ladder itself. A. If you have the ladder extended 15 feet to work on a window, how far should the base of the ladder be from the wall? B. How much farther should the base of the ladder be moved if you extend the ladder to 25 feet to hang Christmas lights?

12. **Walk with me, my friend:** Your school is on the corner of a block, and your home is very nearly on the opposite corner of the block. Normally, to go home, you walk diagonally through a park 509 yards. Your friend lives on the corner of the same block between your home and the school and asks you stop by to visit before going home. Your friend walks 120 yards to get home from school. How much farther will you be walking today to get home?

13. **I forgot my combination:** It's the first day of spring, and you'd like to ride your bike to school. Your bike lock is closed, and you have forgotten the combination (really a permutation). If the bike lock has three dials with the digits 0-9 each, A. How many possibilities are there if the code cannot have all three digits the same? B. If you could try one new code every 5 seconds, how long would it take you in minutes to try all the combinations? C. Do you think you should get a new bike lock? D. How many possibilities would there be if the code does not have any numbers that repeat?

14. **Pay up:** You are going on your senior trip to Myrtle Beach from Detroit! You are going with three other people in your Jeep, which averages 21 miles per gallon.

A. If gas is $2.29 per gallon, how much will each person owe, if split equally, in gas money? B. If your average speed is 55 mph, how long will it take you to get there?

15. **Math work in yard work:** You are helping your dad landscape the backyard. Dad has decided that around two trees, to cover the roots, he would like to put 6-inch long red bricks around the tree about 3 feet from the base of each tree. Each tree base has a diameter of about 2 feet 10 inches. A. How many bricks will you need to put around both trees in total? Your dad also wants to fill each circle with pea pebbles about 2 inches deep. Each bag of pea pebbles holds .5 cubic feet.
B. How many bags will you need in total to purchase?

16. **Skateboard ramp:** You are building a skateboarding ramp for your driveway or sidewalk. You have an 8-foot-long board for your ramp, and the instructions say your angle to the ground should be no more than 24 degrees. At what height should you start building the ramp?

17. **Chemistry class:** In chemistry class, you need to use 2 liters of a 30% acid solution for your experiment. When you go to get the solution, you find your classmates have used it all. Your teacher suggests you combine a 50% acid solution and a 20% acid that are available. How much of each solution should you use to get your 2 liters of 30% solution?

18. **Puppy pen:** Congratulations! You just got a 2 new puppies. Because your backyard has no fencing, you decide to build a rectangular pen from 30 feet of fencing you have in your garage to keep the puppies safe while playing outside. A. What is the area of the largest rectangular pen you can make so your puppies have lots of room to play using only and all your fencing?
B. What shape would you make the pen to make the

largest area possible, and what would that area be?
C. How could you make a larger rectangular pen using the fencing you have?

19. **Cross country race:** You are running in a 5-mile cross country race, and you want to win first place! You are generally the fastest runner on your team, so you want to beat the fastest runner on the opposing team, of course. You check the statistics available and find the best runner on the opposing team has an average speed for the season of 7.75 minutes per mile. If the opponent maintains this average for this race, and you run an average of 7 minutes per mile for the first 2 miles, what does your average speed need to be for the last 3 miles to beat your opponent?

20. **Student council president poll:** You are running for student council president against one other formidable opponent. Your campaign committee decides to take a lunchtime poll two days before the election to check your standing. 104 students said they would vote for you, and 96 students said they would vote for your opponent. Using a ±3% confidence level, can you say decisively that if the election took place today that you would win?

21. **Pop Spanish quiz:** Oh no! You have a pop Spanish vocabulary quiz in front of you, and you have not even done your vocabulary assignment yet! No problem, right? It's only 5 true or false questions. A. If you randomly guess on each question, what is the probability you will get all the problems correct? B. What is the probability you will pass with a B (80%) or better?

22. **The big purchase:** You have been saving your earnings from your part-time job at the grocery store to buy a big screen TV for your bedroom (to watch math videos, of course). Bob's TV Warehouse has a sale on a

50" TV for 20% off the original price of $424. Carlson's Electronic Market has the same TV with an original price of $410 with 10% off the price, and you have a $10 coupon to use on the sale price. Which store has the better deal?

23. **Rollercoaster wait:** You are the 295th person in line to ride the new Lightening Leopard rollercoaster. Each rollercoaster train holds 40 people and a new train leaves every 3.5 minutes. How long will your wait be?

24. **Saving is always important:** You graduate college and at the age of 25 decide to start saving $300 at the end of each month into mutual funds. A. How much will you have in savings at the age of 55 if you earn an average of 8% per year? Let's say you wait for 10 years to start saving (at the age of 35), still at $300 per month with an average rate of return of 8%. B. How much will you have in savings when you are 55? C. What is the difference in savings?

25. **Gym membership:** You make a New Year's resolution to get in better shape, so you decide to join a gym. You have two different gyms very close to you. Jeff's Gym offers a membership for no upfront fee and with a $25 monthly fee. Smith's Gym has a $50 signup fee, but only charges $20 per month. If you plan on joining the gym to get in shape for summer, A. Which gym should you join to pay the least and B. How much would you save over the 2nd gym (summer is in 6 months)? If you plan to work out the entire year, C. Which gym should you join for the least cost and D. How much would you save over the 2nd gym?

26. **Your first new car!:** Congratulations! You are a senior and purchasing your first new car using your part-time job savings. Your particular model of car loses a

standard 17% of its value each year from the year before (depreciation) over a 10-year period. At the end of 5 years, what is the remaining value of your car as a percent from its original value?

27. **And..your first new car loan:** Congratulations! You are a senior and getting your first car ($13,000 and you are putting $1000 down) to commute to college in the fall, and with it, your first car loan. There are great deals going on at the dealership, and you need to know which deal is lowest in cost. In the first deal, you get a $2000 rebate and your loan interest rate is 7%. In the second deal, there is no rebate, but your interest rate is 0%. If your loans would be for 60 months, A. Which deal has the lowest total cost over 5 years? B. What is the difference in cost?

28. **Incredible soccer player:** Suppose you have an amazing record on the soccer field of scoring a goal 96% of the time you take the shot. What is the probability if you take 10 shots in one game, that you will make all 10 shots?

29. **Entrepreneurship:** You decide to start your own business mowing lawns in your local neighborhoods. You buy a push lawn mower for $205.00 and a weed whacker for $65.00. If your average charge per lawn is $16, and you mow 5 lawns per week, how many weeks will it take you to pay off your equipment?

30. **Chess team captain:** As captain of the chess team, you must schedule your 11 players (and you) to play each other 3 times during the season. How many games must you schedule?

31. **Halloween candy:** You are purchasing candy for a Halloween party at your home, and you'd like to buy your best friend Chocolate Blasters, his favorite bag of candy. What is the better deal? A bag of Chocolate Blasters with 50% more candy in it with no price change, or a bag of Chocolate Blasters that is 50% off in price with no size change, or is the value the same?

32. **Watch out for children playing ball:** A fairly accurate, simple formula for stopping distance on dry pavement in feet (y) based on the speed of a car in miles per hour (x) is $y=(1/20)x^2 + x$. A. You are driving through a subdivision with a standard speed limit of 25 mph when a basketball rolls in the road about 20 yards in front of you. What is your total stopping distance if you are going the speed limit? B. Will you stop before the ball? C. How much farther will your stopping distance be if you are going 40 mph? D. Will you stop before the ball? E. A similar formula for stopping distance(y) for an auto on wet pavement going a certain speed in mph (x) is $y=(1/10)x^2 + x$. If you are going the speed limit through a subdivision (25 mph) how much farther is your total stopping distance on wet pavement than dry pavement at this speed? F. Will you stop before the ball going the speed limit on wet pavement?

33. **The power of giving starting with one person:** In the great social media-fueled ALS Ice Bucket Challenge of Summer 2014, a person would be challenged to either have a bucket of ice cold water dumped on them while donating $10 to ALS (Amyotrophic Lateral Sclerosis) within 24 hours of the challenge or to simply donate $100 to ALS Research. Each person who accepts the ice bucket challenge nominates 3 other people for the challenge. Typically, two of the three people accept the Ice Bucket Challenge (with each nominating 3 other people) and the 3rd person donates $100 only. How

much money would typically be raised in just one week due to your accepting the challenge?

34. **Canoe trip:** You and your family are vacationing on the Platte River in Northern Michigan. You decide to go for a canoe ride, and your mom requests that you are back within two hours for lunch. The river flows at 2 mph. If you travel 45 minutes downstream paddling steadily at 6 mph, A. Will you make it back home in time for lunch paddling at the same pace, and how long will the total trip take? B. What would your pace need to be on the return trip to make it back in exactly 2 hours?

35. **A candy gift box:** You would like to fill a box with all of your little sister's favorite kinds of candy for her birthday. You find a box that is about the size you'd like: 10 inches tall, 6 inches wide and 14 inches long. You have some really cool wrapping paper with her favorite animal on it -- cats -- in birthday hats, and you see you have about 5 square feet of wrapping paper. Do you have enough wrapping paper to cover the box?

36. **Social media – be careful what you post:** You post the cutest video you have ever taken of your Beagle Bruce (wearing a tutu and dancing) on Facebook to show your 256 friends. One half of these friends each share your picture with about 64 peeps. Half of these peeps share with about 16 individuals. Half of those individuals share with about 4 humans. And finally half of those humans share with 1 person. A. How many total people have seen your Brucie? B. If this sharing took place over 3 hours, how many people saw a video of Brucie each second?

37. **How math is used in history:** You currently have an 87% in your history class. You would like to have an A (minimum 90%) as your final grade. If your final exam is

worth 20% of your total grade, what percentage do you need to earn on your final exam to attain an A, and is it possible?

38. **Filling the backyard pool:** Now that your younger brother is old enough to swim well, your parents put up a cylindrical above-ground pool in your backyard for Memorial Weekend that is 24 feet across and 52 inches deep. Your parents ask you to fill it using your average size garden hose. About how long will it take you to fill the pool in hours?

39. **High tide and low tide:** You and your family have rented a home on Virginia Beach. You haul your chairs, coolers and umbrellas down to the water around 1 pm at low tide and set everything up about 18 feet from the water. You notice the grade of the beach sloping toward the water is about 15%. Clouds roll in, and you all decide to go shopping and dining, and then you return to the beach about 7 pm when the high tide occurs. About how far are your belongings from the edge of the water now?

40. **Your first aquarium:** You are given an aquarium that is a standard size of 27 in long by 12 in wide by 16 in tall. You should fill the aquarium one inch from the top. You would to like know how many Neon Tetras your aquarium can safely hold so the fish can live in a healthy environment – most new fish owners add too many fish to a tank. The assistant at the pet store says that each Tetra requires 1 gallon of water of space. What is the maximum number of Tetras you could put in your tank safely?

41. **Math National Honor Society fundraiser:** You are a member of your school's Math National Honor Society, and as part of your service hours, you will be running a

game at their annual fundraising event to purchase math story books for elementary school students. The game you make has 10 rubber duckies floating in a pool each with a number from 1-10 on their bottoms. A player pays a certain amount to play, then names a number from 1-10 and chooses a ducky. If the player chooses the ducky with the number they named, they win a prize that is worth $5. You would like your profit to be on average $.40 per play. What should you charge each player to achieve this profit?

42. **Gambling was invented to make a profit from you:** Congratulations! You turned 18 years old and decide for the first time to spend the $2 minimum on a PowerBall ticket. What is your expected return each time you buy a PowerBall ticket if the pot is at the $40 million minimum when you play, just taking the jackpot amount into consideration – not other prizes?

43. **Running your workout:** You would like your twice-weekly workout to include a 3.5-mile run. You have access to your high school running track. To help you keep track of the number of times you circle the track and to break up the monotony, you start in the inside lane and each time you make a full lap, you switch to the next closest outside lane. You follow this method from the inside lane to the outside lane and back to the inside lane. A. Have you completed at least a 3.5-mile run? B. How far have you run exactly?

44. **Cellphones and driving:** According to most recent data, 23% of all car accidents involve cellphone use. If this percentage was true currently, A. Approximately how many of the reported traffic fatalities of 35,200 in the US involved cellphone use? 35% of teenagers admit to texting while driving, and 25% of teens respond to at least one text each time they drive. If the average time

answering a text message is 5 seconds, and you are traveling at just 42 mph, B. How far did you travel in feet while distracted? C. How does this length compare to the length of a football field?

45. **Get more education after high school...no really...seriously...do it:** Recent statistics show that the median weekly earnings for a high school graduate is much lower than the median weekly earnings for a person with a master's degree. A. If the high school graduate (and not beyond) works from the age of 18-65, and the master's degree holder works from ages 23-65, who earned more, and what was the total difference, even though the master's holder worked less years? Statistics also show the unemployment rate for persons with a high school diploma (and not beyond) is greater than the unemployment rate for that of holders of master's degrees. B. Using recent statistics, how many times is the high school diploma holder more likely to be unemployed than the person with the master's degree?

46. **Fun at the casino...right?** Your friends take you to the casino for your 18th birthday, and you decide to play the $1 slot machines. You did some research and found slot machines are normally set up for a 97% return to the player and 3% to the casino. Not bad, right? A. If you have a $100 to spend, after how many pulls on the slot machine will you be most likely left with $1? B. If you pulled the slot handle every 30 seconds, about how long did it take for you to be left with only $1 in your hand? C. How much money did you lose per minute?

47. **Common ground:** You are struggling with a small math concept on probability and want to understand it before you take your quiz tomorrow. Your teacher would like to help you, but has to coach a track meet after school today. Your teacher says she has 20 minutes free

to wait and/or help you sometime between 4 and 5 pm randomly – whenever the track meet finishes. You can only wait and/or get help for 20 minutes anyway, because you have to see other teachers also for other assistance, and so you will also arrive at your teacher's room randomly between 4 and 5 pm. What is the chance you and your teacher will be at her room at the same time at all?

48. **Big lake swimming:** Your friend lives directly north across Big Platte Lake from you. Big Platte has an unusually fast flow rate for Northern Michigan Lakes and actually has a current of about .5 mph running west. If you attempt to swim the 1.6-mile width of the lake north at your average speed of 2 mph, A. How far east will you have to walk once you hit shore to be at your friend's cabin? B. How far do you actually swim? C. If you want to swim directly to her house, about what bearing from north should you take?

49. **Gas mileage and your carbon footprint:** You currently have a big old car that your parents gave you that gets about 17 miles per gallon. You currently drive about 200 miles a week and gas is about $2.39 per gallon. You are considering getting a car with better gas mileage. A. If you purchased a car that gets 40 mpg, how much money would you save each month on gas? B. How many less pounds of carbon dioxide (and other harmful environmental pollutants) will be produced in that month? C. If you purchased a car using biofuel, which produces only 67% of the carbon emissions of a vehicle that uses regular fuel, how many pounds of carbon emissions would be saved each month by buying a 40 mpg biofuel car instead of a 40 mpg regular fuel car?

50. **Wigs for children:** You would like to grow out your short hair so that you can donate it for wigs for children. Your donated hair needs to be at least 10 inches long to make a hairpiece. If you will grow your hair out and cut it off to the same length it is now, how many months will it take you to be able to donate your hair?

51. **Easy weight formula:** Supposedly, a fairly accurate formula (without taking gender into account) of the normal weight of a person between 60 - 80 inches (x) tall is weight=5.5x - 220. If your height is between 60 - 80 inches, put your height in inches into the formula.
A. What weight does this formula produce for you?
B. Does the weight the formula produces seem reasonable or accurate?

52. **Monument Valley rock formation…amazing!:** After you came back from a vacation out West, you print a gorgeous picture of yourself standing next to an amazing rock formation (which took you 3 hours to climb to see!) When you're showing the friend your picture, she asked about how tall the rock arch was. If you are 5 feet 7 inches tall and in the picture you are 12 mm tall, about how many feet tall is the formation if in the picture it is 133 mm tall?

53. **Interior decorating:** Your parents agreed to let you paint your own bedroom any color you wish if you buy the paint and do the painting (carefully!) Your room has a standard wall height. It is 10 feet by 11 feet with a nook that is 6 ft by 4 ft. Your closet door and door to the room are both 2 ft 8 in by 6 ft 8 in. Your window is 3 ft by 6 ft. You will need to use two coats of paint, but will not have to paint the ceiling. How many gallons of paint do you need to buy?

54. **Gardening is fun:** You and your family really enjoy tending a garden in the warmer months and the food it produces. As a family project, you decide to build a greenhouse so you can continue gardening in the colder months. Your grandpa decides to build a dome-shaped greenhouse with a 20 foot by 14 foot rectangular base and an arch height of 7 feet in the shape of a half circle. Your job will be to secure the plastic on the structure over the arch and at both half circle ends.
A. How many square feet of plastic will you need to purchase? Your grandpa says you can use 1/6 of the greenhouse for your own plants. B. What volume of space will you have available to you?

55. **World population:** A. Approximately how many people are added to the earth's population each year? About only 43% of the world's land is habitable. B. How many square miles are available per person currently? C. If the world population continues to grow at the current pace, how many years from now will each person's available space equal about only 100 ft x 100 ft --10,000 square feet?

56. **Frugal eating:** It's dinner time, and you have a taste for pizza and snow cones...and you want the best value. You go to the new pizza parlor in town that has an ice cream parlor too. Since you plan to come often, you'd like to know what the best deals are. For your pizza, you'd like cheese, pepperoni, bacon and onion. A 12-inch medium round cheese pizza is $8.99 with $1.00 per topping. A 16-inch large round cheese pizza is $9.99 with $1.50 per topping. A. Which pizza is the better deal for you? You then head next door to the ice cream shop for snow cones. The small snow cone is 6 inches tall and 3 inches in diameter and is $1.50. The medium snow cone is 8 inches tall and 4 inches in diameter and $3.50.
B. Which snow cone is the better deal?

57. **Meeting with STEM friends:** At a STEM conference you made two friends. Your parents agreed that you can get together and visit during the summer. You feel it would be fair for each of you to travel the same distance to meet, plus it would be fun to visit a new city together. A. In approximately what larger city should you meet if you each live close to Milwaukee, WI and Las Cruces, NM and Bismarck, ND? B. Approximately, what is the average distance each of you will travel?

58. **I give you credit for knowing this:** For (very soon) future reference, when you go off to college, credit card companies will hound you to get a card with them. Two reasons: 1. Even if you don't have much income now, you will soon. 2. They know you don't know much about credit cards. Let's say you take a credit card deal -- you are offered a $1,000 credit line at 18.3%. And, let's also say, you max out your card very quickly (it happens often) buying cool stuff for your dorm room, new clothes, going out to eat and buying concert tickets. You have no more room on your credit card, but now have payments to make. You cut up your credit card and vow to pay the balance off. A. If you pay the minimum of $25 per month, how long will it take you to pay off the credit card? B. When you are done paying it off, how much interest have you paid in total?

59. **Why did I forget everything I learned in school?:** Did you know that when you are tested on facts (that does not have direct relevancy or use in your daily life) and then you are retested on these facts (without review) a certain number of "t" months later, that a logarithmic equation fairly closely models the human retention rate? The equation (assuming 100% memorization at time zero) is $Score(t)=100-30\log(t+1)$. A. If you scored 100% on your Spanish vocabulary memorization test at the end of the

course, and then did not use Spanish in the next year, what would be your retention rate approximately after 12 months? B. Two years?

60. **That didn't help me:** You took penicillin for the first time in your life to help you overcome a bacterial infection. Unfortunately, it made you feel very itchy and uncomfortable. How long will it take until the dose is 99% out of your system, if the half-life of penicillin is 1 hour?

61. **Future Leaders in Law conference:** Your school is offering a fieldtrip to a law conference you are very interested in attending. The cost depends on the number of students attending because it costs $650 to register your school at the conference and then $40 per student. Currently, 58 other students are signed up. How many more students do you have to convince to go if you would like your cost to not be over $50?

62. **Charity talent show:** You are in charge of ticket sales for your high school's talent show, to raise money for homeless families in your community. You look at the past years' ticket sales. You found in the first year of the talent show, ticket prices were $3 and 550 students attended. Each year after that, the ticket prices went up one dollar while ticket sales went down about 50 people. A. What price should you charge for tickets to maximize your donations? B. How much money will you most likely collect?

63. **Shake, rattle and roll:** A sizable earthquake occurs with the epicenter close to Fairbanks, Alaska. How long will it take the seismic waves to reach your hometown, given the earthquake is large enough for you to feel the tremors?

64. **Pizza's here:** Your part-time job is as a pizza delivery person. You work for Cheezy Pizza Parlor in the heart of your suburban town, which is approximately 3 miles long by 5 miles wide. You are a little short on gas and cash this evening and are hoping that your last three deliveries are in a 1 mile radius from the shop. What is the probability of that occurring?

65. **I'm going to be a professional athlete:** In your state: A. About how many women high school basketball players will go on to college basketball? B. How many women high school basketball players will go on to professional basketball? C. About how many men high school basketball players will go on to college basketball? D. How many men high school basketball players will go on to professional basketball? E. Need a back-up plan?

66. **Raise us up:** You and your friend work part-time for a summer ice cream stand -- your friend has worked 2 years longer than you. Your friend earns $10 per hour and gets a $1 raise. You earn $8 per hour currently and get a $.90 cent raise. A. Who earned the highest raise per hour? B. Who earned the highest percentage raise?

67. **It's the same...right?** If the price of a $200 bike goes up 30% due to demand, then down 30% at the end of the season, A. Is the price of the bike now $200? B. If not, what is the new price?

68. **What a beautiful...and long sunset:** You are driving west with your family on Route 66 through the Mojave Desert in California in the evening. The sun seems to be hanging in the sky longer than usual -- is it? If you stopped your car and watched the sunset, and the sun would take one hour to set, approximately how much

longer will it take the sun to set for you in minutes if you drive west at 75 mph for one hour?

69. **The dreaded college tuition rates:** For the last 10 years from 2007 to 2017, college tuition rates for 4-year colleges have increased an average of 3.5% per year. The current average college tuition (not including room and board) at state public universities for in-state students is $9,650 per year. A. What does a 4-year degree currently cost in tuition only, if 2017 is your freshman college year? B. If this trend continues, what will be the tuition cost of a 4-year degree for our current 2017-2018 high school freshman if they enter college in the fall of 2022? By the way, median household income in 2016 was $56,156. Also, by the way, the current average tuition at a two-year college is only $3,347.

70. **Getting "mean"ing from data:** You are doing an article for your school newspaper and polling students to find their average weekly earnings before taxes. Of the 200 students you surveyed, 130 of them earned $0. 30 students answered $1-100 per week, 30 answered $101-200 and one answered $2,000 per week because she has owns her own successful website development business. A. What is the mode? B. What is the median? C. What is the mean? D. What is the most accurate way to report this data?

71. **Hard-earned money:** You have a 20-hour-a-week job at which you earn $9.00 an hour. You also have some favorite habits: You go to Starbucks 4 times a week for a $4 latte, and you go the movies with your friends every other week ($8 ticket, $8 large popcorn, $4 soda). A. What is the total cost of your Starbucks and movie habits per year? B. How many work hours and work weeks did you work to pay for these habits?

72. I'm on fire (scary info following): In 2015, 5.9% of teenagers characterize themselves as daily cigarette smokers by the time they leave high school. A. If this is true at your high school, about how many current students at your high school will leave as daily smokers? If the current trend in smoking continues, 1 out of 3 of these seniors will die prematurely (average of 10 years) due to smoking. B. How many teenagers at your school will die prematurely? An average adult smoker smokes 1 pack a day. C. In your state, how much does this habit cost the average smoker per year?

73. Your first pay check...wait...what? You have started your first job and are working 20 hours a week at $9 an hour. After two full weeks, you get your first full paycheck with gross earnings of $360. A. After taxes, what final amount will be on your paycheck? B. About how many hours in those two weeks did you work simply to pay the government?

74. Boy, am I thirsty: Lake Mead supplies Las Vegas with 90% of its water supply. In 1998, Lake Mead was at its highest level at 1,296 ft. In 2017, Lake Mead is at only 40% of its capacity at a level of 1,079 feet. A. If the surface area of Lake Mead is about 247 square miles, what loss in gallons is that on average per year during that time period? B. The pipes that feed water to Lake Mead are 227 feet below the current water level. At the current rate of water loss, when will the water level fall below the pipes? C. An emergency pipe can be opened at great risk and great cost that is 190 feet below the original pipes. How many more years of water use will this emergency pipe provide to Las Vegas at the current water loss rate? In Las Vegas, the average consumption of water is 125 gallons per person per day, of which 70% is used for lawns, parks and golf courses.

D. How many gallons of water is used per year in Las Vegas simply to keep grass green?

75. **Be kind to each other:** You are sitting in a health class of 30 students, and your teacher gives the following percentages to you. For each of the following categories and percentages, find how many fellow students in your class would be in each category:
A. 12.5% of minors will lose a parent or sibling. B. 13% are learning disabled. C. 7% have a chronic physical illness D. 21% suffer from mental illness E. 67% experience a traumatic event by age 16 F. 22% live in poverty. So, the real question is G. What percent of students in your classroom should you be kind and helpful to always?

76. **Annual city art show:** You are in a contest in your city's annual art show and have entered your painting in your age category. There are three judges. Each judge awards points in the following way: 1 point for minimum demonstration, 2 points for good demonstration and 3 points for excellent demonstration. The 3 categories are Creativity/Originality (50% weighting), Composition/Design (30%) and Craftsmanship/Skill (20%). You see the scores posted (see chart), but the scores/winners are not posted yet for 1st, 2nd, 3rd place, so you determine the winners before the judges do. Who won 1st, 2nd, and 3rd place?

	Creativity	Composition/Design	Craftsmanship
You	8	3	7
Butch	5	9	4
Carl	9	6	3
Darlene	8	5	9
Evan	4	8	7

77. **Stay ahead of the game:** Current data shows that each time a high school football player engages in a practice or a game, that player has a 76.8/100,000 chance of getting a concussion. A. What is the probability of a player getting a concussion during one full season? 33% of high school football players with one reported concussion in a season have two or more concussions. B. What percent of all high school football players will have two or more concussions in one season? C. Approximately how many high school football players in the United States suffer from one or more concussions each season?

78. **Java jitters:** For young adults between the ages of 13-18, a maximum of 100 mg of caffeine per day is encouraged as a limit. A. How many Monster Energy drinks (smallest-size, regular) can you drink in one day to hit this maximum? B. How many 12-oz cans of coke is equivalent to 100 mg of caffeine? C. How many Starbuck's Grande brewed coffees can you drink in one day?

79. **Water in the sky (not rain):** Water towers are used to provide water in emergencies and during peak times of use. Your city has a water tower, and the city requires 75 psi of water pressure on average. A. At least how tall must this water tower be if it is placed on a hill 20 feet high? B. If there are 13,000 households in your city, and the water tower is designed to provide water for a full day to all households in an emergency, about how many gallons of water does it hold?

80. **Saving a life:** You are standing on a secluded Lake Michigan beach without lifeguards. You are standing at the shoreline and see that someone is calling for help about ¼ mile out in the water and ½ mile down the shoreline, and they need you to save them. You can run

at 12 mph and can swim half as fast as you can run. Time is of the utmost importance. A. To have the shortest time of rescue, how far do you run down the shoreline in feet and miles before you dive in the water and swim directly to this person? B. What is the shortest time in minutes of rescue?

81. **"Wood" you like to build a playhouse?:** You would like to build your toddler niece a wooden playhouse made of 2 X 4 pieces of knotty pine wood. The playhouse will be 4 feet wide by 5 feet long by 5 feet tall. You would like to place a 2 feet by 2 feet window 1 foot from the top of a wall exactly in the middle on one of the larger walls. You would like a door 4 feet tall by 2 feet wide door in the other larger wall also in the middle. Each length-wise space will be filled with be one board – not filled with pieces of boards. The roof will be made of metal. Should you buy 2 inch x 4 inch x 8 feet boards or 2 inch x 4 inch x 10 feet boards to buy the least total feet and have the least total waste? Both types of boards are the same cost per foot.

82. **Fast growing virus:** You are a medical researcher studying a new case of a fast growing flu virus in your state. The number of cases reported are growing each week. Last week there were 98,415 cases reported, and this week a startling 295,245 new cases. You are trying to determine how many weeks ago this virus was first introduced to the population and by how many people to help understand how and where it began so you can halt it. A. How many weeks ago did this virus start in the population and B. Approximately how many people first got the virus?

83. **Dimensions are important:** You are buying your favorite thin crust pizza at your local pizza place. You always order the square foot-long pizza. New item alert!

The pizza place is offering the same pizza but with 2-foot sides for double the price of the original. A. Are both pizzas equal deals in price or is one pizza less expensive than the other per square foot? You are buying a box full of chocolates for your grandma for her birthday. There is one cubed box of chocolates for $10 and one cubed box of chocolates whose sides are twice as long for $20. Your grandma would want you to get the best deal. B. Are both boxes the same value or is one box a better value than the other per unit cubed?

84. **Time is of the utmost importance:** You and your friend have spent hours in the park just walking the paths and talking. Dusk is approaching and your friend suddenly realizes in a panic that his phone is missing.
A. If you are in Park A, can you walk the entire pathway with only taking each path once for efficient searching?
B. Can you add one path to Park A to make it "efficient" – called traversable? See paths of park below:

85. **It's prom time!:** You are the head of the prom committee, and it is now your job to ensure that prom ticket sales during lunch periods for the next two days run smoothly and quickly. You are expecting 325 ticket purchasers (customers) over the next 6 half-hour lunch periods, and you expect them to arrive to your table steadily. Each student council member helping at the ticket table can take care of a total transaction (take the payment, write receipt, choose meal and seat) for a buyer in about 7 minutes. If you would like customers to

wait on average 10 minutes in line or less, how many student council members do you need to help sell tickets?

86. **Going, going, gone?:** You have been trying all season to get a home run on your standard high school's baseball field with the required 6 feet high fence around it. You have done your research and practice, and you know the best angle to hit the ball is about 45 degrees, and you normally connect with the ball 5 feet from the ground. You have gotten your exit speed (the rate the ball comes off the bat) up to about 98 mph on your best swing. You know air resistance usually cuts the length the ball travels in half from calculations from physics formulas. A. If you aim for the closest wall from home plate, will you get a home run under these conditions? B. If so, by how many feet did the ball clear the fence?

87. **Just once?:** A. Find two 3-digit numbers whose sum is a 3-digit number and each integer 1-9 is used only once in the problem and solution. B. Is there only one possible solution?

88. **How does it stay in the air?:** If a typical cumulus cloud is about 1 kilometer in length, width and height, and the typical density of a cloud is .5 grams of water per cubic meter, how much does the typical cloud weigh in kilograms and pounds?

89. **Not as much as you would think:** If a woman from the United States of average height could walk all the way around the earth at the equator, how much farther would her head travel than her feet? Hint: You don't need any info about the earth to solve this problem!

90. **Tick tock:** What is the degree measure of the acute angle formed by the hands of a 12-hour clock at 1:26?

91. **Make an equation:** A. 1 2 3 4 5 6 7 8 9 0 = 100. Leaving all digits in order, use + or - signs to make the equation true. You may connect two values to form 2-digit numbers. B. Is more than 1 solution possible?

92. **Start small:** What number is in the ones or units place when 3^{1000} is calculated?

93. **Rush hour or not:** You drive 40 miles from your house to your grandma's house to visit her on her birthday at an average speed of 40 mph. On your way home, it takes you twice as long due to rush hour. What was your average speed for the entire trip?

94. **Whose grade is correct?:** You took a multiple choice quiz that has ten questions each worth 5 points. 5 points is awarded for a correct answer, only 1 point is awarded for a blank answer (student recognizes they do not know the answer at least) and 0 points for an incorrect answer. You receive your quiz back and your score is 43 out of 50 points. How do you know for sure your teacher graded this quiz incorrectly?

95. **Multiplying fun:** A. Using each of the digits 1 through 6 only once, write a true equation: ? X ??=???. B. Is there only one possible solution?

96. **What's the better deal?:** Congratulations! You have won your country's science competition where you have created a method to reduce all types of pollution by 50% virtually cost-free! You are offered your reward over a period of 30 days to further develop your project. The contest coordinators allow you to choose your reward from two options. Option 1: You receive $100,000 immediately. Then, on the first day of the month and then each day after, you get paid $5,000 more than you

did the day before.   Option 2: Receive $.01 immediately. Then, on the first day, you get $.02 and each day after you are paid double your pay from the previous day up to day 30. A. How much would you get paid with Option 1 on the last day of your reward (30 days)? B. With Option 2? C. Between what two consecutive days did the pay of one option become better than the other?

97. **Head or tails?:** Your friend says that the probability of 3 thrown dice turning up all evens is 50% or all odds is 50%. He says the reason is that when you throw three dice, two of the dice MUST match (odds or evens), so the third dice -- which has a 50% chance of being odds or evens – is the only dice that make the difference in the outcome and so the probability also. A. Is your friend right? B. If not, what is the probability of three thrown dice turning up all evens? C. Or all odds? D. What is the probability of all three dice being all even or all odd?

98. **Amazing facts of YOUR life:** If you live an average lifespan for your gender from the day you are born: A. How many times will your heart beat if you use an average normal heart rate? B. How many seconds will you have lived? C. About how many pounds of food will you have eaten? D. How many pounds of trash will you have produced? E. How many hours will you have slept?   F. Find another fact you would be interested in knowing, such as these, and find your data and make your calculation!

99. **Easy add:** What is the sum of the first 1000 integers?

100. **I need my space:** New York City, NY has the highest population of any city in the US. A. What is the population density of New York City in square feet per person? B. What is the most densely populated city in

the world? C. What is the population density of this city in square feet per person? D. What is the size of your classroom in square feet? E. How many times denser is the world's most populated city versus New York City? F. What is the population density of your city in square feet per person, and how many times less dense is it than NYC?

101. **Paid to go to school?:** According to recent data, a person with a high school diploma (and no more) will earn $272,000 in their lifetime more than a person with no high school diploma. Based on this information,
A. About how much are you being paid per hour to go to high school, or how much is each hour worth, if you finish high school and receive your diploma than if you did not? B. Are you working hard while in school to earn your hourly pay?

102. **Number triangle:** A. Arrange the numbers 1 through 6 in a triangle so that the sum of each straight line of three numbers gives the same sum. B. Are there any more possible arrangements? C. How does your solution here compare to the solution in #95?

```
      X
     X  X
    X  X  X
```

103. **Hotdog House:** You see a commercial for The Hotdog House Restaurant right next to your neighborhood. The commercial claims they sold 3.5 million hotdogs last year. The restaurant has one location and has 6 full-time employees. The restaurant is extremely busy at all times. Is it mathematically possible that this restaurant sold 3.5 million hotdogs last year? Decide if this claim is true or false, and use math to support your answer.

104. **Are you sure I can't use a ten?:** If X and Y are both positive integers and neither is a multiple of ten, determine X and Y, if X times y equals 10,000.

105. **Why do we need letters in math?:** What are the numeric values associated with the letters F, G, H, J, K if each letter has a different value and F G H J H −J K F F H = G F J F J?

106. **Party punch:** You have enough drink mix to make exactly 2 gallons of lemonade for your campsite party. You can only find an 8-gallon bucket and a 5-gallon bucket to help with measuring. How can you use these two buckets to measure out exactly 2 gallons of water to make your lemonade?

107. **Big winner every time:** You are playing a game with your friend. You set 35 bingo chips in a row. On each turn, you may choose 1,2,3 or 4 chips. The person who is forced to take the last bingo chip is the loser. You go first. What method should you use to win the game every time?

108. **Nines only:** A. Using only single-digit nines and the operations of addition, subtraction, multiplication and division, write an equation that equals 100. B. Is there more than one solution?

109. **This is no phone-y:** Write the first 3 digits of your phone number without the area code. Multiply by 80, add 1 and multiply by 250. Add the last 4 digits of your phone number and add them again. Subtract 250 and then divide that number by 2. A. What is your solution? B. Why does this math work in this way?

110. **Choo choo!:** Suppose a train is traveling at 75 mph when you see it enter a mountain tunnel. It takes 5 seconds for the train to completely enter the tunnel. It then takes 40 seconds for the train to completely pass through the tunnel. A. How long is the train in feet? B. How long is the tunnel in feet?

111. **Cake for everyone!:** You have baked a round cake for your mom's birthday. Using only 3 straight cuts all the way across the cake, what is the most number of pieces you can make?

112. **The Great Pyramid of Giza:** This pyramid was the largest structure in the world for about 3,800 years. If each block composing the structure is on average 1.5 meters x 1.5 meters wide x 2.25 meters deep, A. About how many blocks make up the pyramid? B. If each block weighed on average 9.3 tons, what is the total weight of the pyramid? C. About how many tons were added to the pyramid each year it was being built? D. About how many pounds were added to the pyramid per slave?

113. **Shuffling a standard deck of cards:** A. How many arrangements are possible for a deck of cards when shuffled? Let's say that with the current world population, each person began shuffling a deck of cards each second of their life for their whole life. B. How many shuffles would have been accomplished? C. Approximately what percent of the total arrangements possible in a deck of cards did the population shuffle?

114. **Incredible salaries:** A. If LeBron James plays all regular season games in the 2016-2017 season, how much does he earn per minute during these games? What if you take into consideration that James practices

40 hours per week, in addition to his games, B. What is his pay per minute for this season?

115. **Wind turbines:** Each wind turbine produces about 3.2 million KWH (kilowatt hours) per year and each U.S. household uses about 10,000 KWH per year. About how many wind turbines would provide the energy necessary to power all the households in the U.S.?

116. **How old am I?...said the tree:** Did you know you can tell the approximate age of a tree if you know its type? This can be done by finding the diameter of a tree 54 inches from its base and multiplying by the tree's growth factor: A. You have a white oak tree in your back yard. You measure the circumference (54 inches from the base) of your tree and find it is 63 inches. A. Approximately how old is your white oak tree? You can also tell how tall your white oak tree is! Did you know that oak trees are generally 16 times their circumference in height? B. About how tall is your oak tree in feet?

117. **Don't squish me!:** The spittlebug is the highest jumping animal in the world given its body length. If the average spittlebug is 6 millimeters long, and it can jump a maximum of 563 mm, A. How many times is its jumping distance versus its length? If your jumping height were the same ratio as the spittlebug's, B. How high could you jump in feet? C. How does this height you could jump in feet compare to the average height of a two-story house?

118. **A long time to think:** A. How many times higher are the tallest waterfalls in the world than the Niagara Falls? If you went over the side of these tallest falls (don't worry, you survive unscathed!) B. How long would you have to

think about your situation before you land in the pool below?

119. **I owe...I owe...** If the U.S. started paying off its national debt of approximately $20 trillion by 10% each year, how long would it take for each citizen's part of his debt to be less than $10?

120. **50/50...right?** A. If you flip a coin 50 times, is the probability of getting exactly 25 heads and 25 tails 50%? B. If the probability is not 50%, what is the probability?

121. **Cone shapes:** Did you know that different materials have different "angles of repose?" This means that when you pour a material from above to a horizontal surface, it will form an upside down right cone, and depending on the material, the base angle of the cone will vary naturally. For instance, earth (dirt) has a natural angle of repose of 38 degrees, but wheat's angle of repose is 27 degrees. A freighter is carrying a load of dirt and a separate equal load of wheat, and using a conveyor belt, unloads the dirt and wheat in two separate piles. A. How many times taller is the cone of dirt versus the cone of wheat naturally? B. How many times bigger is the radius of the base of the wheat cone compared to the dirt cone?

122. **Chirp me the temperature:** It is true that you can gauge the temperature outside by the frequency of cricket chirps. You are lying in bed on a summer night, and a cricket is chirping outside your window. You set the timer on your phone for 1 minute and count cricket chirps. You count 128 chirps. What's the approximate temperature in Fahrenheit degrees?

123. **Precious cargo:** A humanitarian aid plane is dropping off a supply crate of food and medicine to a small village of a country in desperate need. It is flying at 150 mph at an altitude of 2,100 feet. After the plane flies over the target: A. How many seconds should it wait to release the package so it lands in the target area?
B. What is the horizontal distance the plane is from the target area at this time?

124. **Do you still need the watch?:** A. How do the G-forces that a typical human being can handle for a short period of time compare to the G-forces a mechanical watch is resistant to normally? B. What is the real question here?

125. **Fermi-type questions for your life:** A. What are Fermi questions? B. Estimate how many pieces of notebook paper are used by students in your school each year. C. Estimate how many tacos are eaten each month by residents of your city. D. Estimate how dentists does your state need to service the residents in a timely manner.

Solution Set

1. a. Need nautical miles between cities: 84.53
b. Law of Cosines or geometric representation with measurement, rate, distance   c. Extra distance: 4.7 nautical miles
d. $a^2 = 50.53^2 + 20^2 - 2 \cdot 50.53 \cdot 20 \cos 32°$, a=35.2 nautical miles so extra distance: 20 + 35.202 − 50.53 = 4.672

2. a. None   b. Solve a smaller problem   c. 79 people
d. 100-20-1=79, smaller problem: A person 10th in line and a person 20th in line have 9 people in between, etc.

3. a. Average growth rate of a blue spruce: 24 inches per year
b. rates, conversion or dimensional analysis   c. 3.7 years
d. (89 inches)(1 year/24 inches)=3.7 years

4. a. Height of student: 5 feet, 6 inches   b. proportions, similar figures, rate, geometric representation with measurement
c. 89.8 feet   d. $\dfrac{3\,ft\,shadow}{5.5\,ft\,height} = \dfrac{49\,ft\,shadow}{X}$, X=(5.5)(4.9/3)

5. a. None   b. rate, distance, conversion or dimensional analysis
c. 9 min running time, so yes, with 1 min to spare
d. $1.8\,miles \cdot \dfrac{1\,hour}{12\,miles} = .15\,min$, (.15hours)(60min/1hour)=9 min

6. a. Speed of sound: 1,125 feet/second   b. rates, distance, conversion or dimensional analysis   c. 2.6 minutes
d. $\dfrac{1,125\,ft}{1\,sec} \cdot 5\,sec = 5,625\,ft$, $5,625\,ft \cdot \dfrac{1\,mile}{5,280\,ft} = 1.06\,miles$,
$1.06\,miles \cdot \dfrac{1\,hour}{25\,miles} = .0426\,hours$ or 2.6 min

7. a. None   b. follow a pattern, solve an equation   c. 6 free throws made   d. $\dfrac{5+x}{16+x} = .5$, 5+x=8+.5x, .5x=3, X=6

8. a. None   b. rates, jobs problem, solve an equation
c. No, you will only get 97.2% of the job done, you need 10.3 hours to complete the job

d. $\dfrac{1\,job}{18\,hours}\cdot 10\,hours + \dfrac{1\,job}{24\,hours}\cdot 10\,hours = .972\,jobs\,completed$

$\dfrac{1\,job}{18\,hours}\cdot x\,hours + \dfrac{1\,job}{24\,hours}\cdot x\,hours = 1\,job$, X=10.3 hours

9. a. None   b. rates, conversion or dimensional analysis
c. 840.3 revolutions/minute
d. $\dfrac{60\,miles}{1\,hour}\cdot\dfrac{5{,}280\,feet}{1\,mile}\cdot\dfrac{1\,hour}{60\,min}\cdot\dfrac{1\,revolution}{2\pi\,feet} = 840.3\,rev/\min$

10. a. None   b. rates, conversion or dimensional analysis
c. No! You need 33 feet and you only have 26-2/3 feet
d. You need: (22 necklaces)(18 inches/necklace)(1 foot/12 inches)=33 feet You have: (20/3 feet/roll)(4 rolls)= 26-2/3 feet

11. a. None   b. Rates, proportions, conversion or dimensional analysis   c. A. 3.75 feet   B. 6.25 feet   d. A. (1/4)(15 feet)=3.75 feet   B. (1/4)(25 feet)-3.75 feet=2.5 feet

12. a. None   b. Pythagorean Theorem, geometric representation with measurement   c. 107.7 yards   d. $x^2 + 120^2 = 509^2$, X=494.652 yards, 120 yards + 494.652 yards – 509 yards=105.7 yards

13. a. None   b. permutation, factorial or solve a smaller problem
c. A. 990 possibilities   B. 82.5 minutes   C. No! Find the code once and write it down!   D. 720 possibilities   d. A. 10^3-10=990
B. (990 possibilities)(5 seconds/possibility)(1 min/60 sec)=82.5 min
D. (10)(9)(8)=720

14. a. Driving distance from Myrtle Beach to Detroit: 796 miles
You may change the $/gallon to reflect current gas prices
b. rates, conversion or dimensional analysis   c. A. $21.41
B. 14.3 hours   d. A. $796\,miles\cdot\dfrac{1\,gallon}{21\,miles} = 37.905\,gallons$,

$37.905\,gallons\cdot\dfrac{\$2.29}{1\,gallon} = \$86.80\,total$, $86.80/4 people=
$21.70/person   B. (796 miles)(1 hour/55miles)=14.5 hours

15. a. None   b. geometric formulas, conversion and dimensional analysis   c. A. 112 bricks   B. 37 bags   d. A. Total radius of tree and from tree to bricks: 17 inches+36inches=53 inches,

$2 \cdot \pi \cdot 53 inches \cdot 2 trees = 666.0 total circumference$, (666.0 inches)(1 brick/6 inches)=111.0029 bricks or 112 (to have extra versus too little).
B. Total area=area from tree to base-area of tree base (twice):
$(\pi \cdot 53^2 - \pi \cdot 17^2) \cdot 2 trees = 15,833.6 inches^2$, (15,833.6inches²)(1 ft²/144inches²)=109.956ft², volume of pea pebbles needed: area x height: (110ft²)(2 inches)(1ft/12in)=18.326 ft³, (18.326 ft³)(1 bag/.5 ft³)=36.7 bags or 37 bags to have extra versus too little.

16. a. None   b. trigonometry or geometric representation by drawing   c. 3.3 feet   d. $\sin 24° = \dfrac{X}{8}$, X=3.3 feet

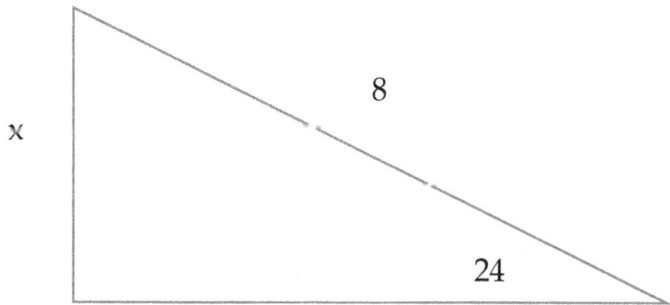

17. a. None   b. Mixtures, linear systems of equations
c. 2/3 liters of 50% solution and 1-1/3 liters of 20% solution
d. x + y = 2 (total solution) and .2x + .5y=.6 (total acid) where (.6=2 liters x 30%), OR -.2x + -.2y = -.4 and .2x + .5y = .6, .3y=.2, y=2/3 liters and then x=1-1/3 liters

18. a. None   b. Unlimited strategies/concepts!: quadratics (with systems), calculus (derivative), logic for shape then dimensions, geometric representations, patterns, charts, use of graphing calculator   c. A. 56.25 ft² (square)   B. circle, 71.6 ft²   C. Use walls of a structure as sides of pen   d. A. Using quadratics: 2x + 2y = 30 or x + y =15 or y = 15-x and maximum area = xy, using substitution max area = x(15-x) or A=-x²-15x, max is located at vertex of this parabola of form f(x)=ax² + bx + c using formula vertex=(b/2a,f(b/2a)), x = -15/-2 or 7.5 ft and y=7.5 ft so area =56.25 ft². B. $2\pi r = 30 ft$, so   r=4.775ft, so area= $\pi r^2 = 71.6 ft^2$.

19. a. None   b. Rates   c. 8.25 minutes per mile
d. Your opponent will take a total of (7.75 min/mile)(5 miles) = 38.75 mins to run this race, and your time has to be just less than 38.75 mins, you used (7 min/mile)(2 miles) = 14 mins in the first 2 miles so you have 38.75-14 = 24.75 minutes left to use, 24.75 minutes/3 miles = 8.25 mins/mile you must run for the last 3 miles.

20. a. None   b. percents   c. No!   d. % of votes for you: 104/200 = 52% with a margin of error range 49%-55% % of votes against you: 96/200 = 48% with a margin of error range 45%-51%, so no.

21. a. None   b. Make a pattern, binomial theorem/combinations and/or use of graphing calculator, Fundamental Counting Principal
c. A. 3.1%   B. 18.8%   d. Using Binomial Theorem (T=true probability and F=false probability and C=combination) where
$_5C_0T^5 + {_5C_1}T^4F + {_5C_2}T^3F^2 + {_5C_3}T^2F^3 + {_5C_4}TF^4 + {_5C_5}F^5$ or
$T^5 + 5T^4F + 10T^3F^2 + 10T^2F^3 + 5TF^4 + F^5$ A. $(1/2)^5$ =3.125% or 3.1%
B. $T^5 + 5T^4F = 3.125\% + (5)(1/2)^4(1/2)$, 3.125% + 15.625%=18.8%

22. a. None   b. percentages   c. Bob's   d. Bob's: $424 x 80% = $339.20 and Carlson's: $410 x 90% - $10=$359.00

23. a. None   b. Conversion or dimensional analysis
c. 28 minutes   d. (295 people)(1 train/40 people)=7.3750 or the 8th train, (8th train)(3.5minutes/train) = 28 minutes

24. a. None   b. Formula or TVM solver on graphing calculator
c. A. $447,107.83   B. $176,706.12   C. $270,401.71
d. Using TVM Solver: A. N=360, I%=8, PV=0, PMT=-$300, P/Y and C/Y=12, solve for FV (alpha enter): $447,107.83   B. Change N=240, $176,706.12   C. $447,107.83 - $176,706.12 = $270,401.71

25. a. None   b. systems of equations, chart or pattern
c. A. Jeff's is cheaper up to 10 months   B. $20 difference
C. Smith's is cheaper after 10 months   D. $10 difference
d. A. Jeff's Gym Cost=25m (m is months) and Smith's Gym Cost=20m + 50, 25m=20m + 50, m=10. Jeff's Gym is cheaper before 10 months when comparing y-intercepts and slopes. B. (20(6) + 50)-25(6) = $20   C. Smith's Gym when comparing y-intercepts and slope   D. (25)(12)-((20(12) + 50)) = $10

26. a. None   b. rates, exponential functions   c. 39.4%
d. Remaining % value each year = 83%, $(.83)^5$ = 39.4%

27. a. None   b. rates, financial formulas or use of TVM solver
c. A. Rebate deal is best   B. $118.80   d. Use of TVM Solver:
A. Cost of rebate deal: N=60, I%=7, PV=-10,000 (13,000-1,000 down-2,000 rebate) FV=0 (alpha enter on PMT) and PMT=$198.01. Total payments over 5 years would be $198.01 x 60= $11,880.60, so interest paid must be $11,880.60 - $10,000 = $1,880.60. So total cost of rebate deal is $10,000 cost + $1,000 down + $1,880.60 interest = $12,880.60. Total cost of no rebate, but 0% interest is $13,000.00.
B. Difference in cost $13,000.00-$12,880.60 = $118.80.

28. a. None   b. Fundamental Counting Principal with probability, exponential functions   c. 66.6%   d. $(.96)^{10}$ = 66.6%

29. a. None   b. rates, systems of equations   c. 3.4
d. Cost=$270 and Revenue = ($16 x 5$)w (where w=week), 270 = 80w, w=3.4

30. a. None   b. chart or patterns, system of equations or regression feature on calculator   c. 198 games (66 x 3)
d. using system of equations and patterns with each players playing another player once, then multiplying by three: 1 player = 0 games and 2 players = 1 game and 3 players = 3 games and 4 players = 6 games, etc. Because the second differences are constant, the pattern of players to games is quadratic: $f(x) = ax^2 + bx + c$. So using the first three sets of data by inputting number of games for f(x) and x = number of players: 0 = a + b +c and 1 = 4a + 2b + c and 3 = 9a + 3b + c, then 1=3a + b and 3 = 8a + 2b, then 1 = 2a, then a = ½, then b = -1/2 and c = 0, so $f(x) = .5x^2 - .5x$. Input x=12 players for x and obtain 66 games. 66 x 3 = 198.

31. a. None   b. rates, conversion or dimensional analysis
c. Best deal is 50% off price with no size change
d. Best deal is the lowest price (P)/original bag size (B): 50% more candy – no price change = 1P/1.5B = .667P/B 50% off price – no size change = .5P/1B, .5 < .667

32. a. None   b. Formula, conversion or dimensional analysis
c. A. 18.8 yards B. yes C. 21.3 yards D. no E. 10.4 yards
F. No   d. A. 25 mph dry pavement: 25^2/20 + 25 = (56.25 feet)(1yard/3 ft) = 18.75 yards, 18.75 < 20 so yes.   C. 40 mph dry pavement: 40^2/20 + 40 = (120 feet)(1 yard/3ft) = 40 yard, 40 > 20 so no. E. 25 mph wet pavement: 25^2/10 + 25 = (87.5 feet)(1 yard/3ft) = 29.167-18.75 yards=10.4   F. 29.167 > 20 so no.

33. a. None   b. develop formula, pattern or chart   c. $15,250.00
d. By Chart:

Day	Total $ for day	Total $ collected for week
0	$10 (you)	$10
1	$120: (2 x $10 + $100)	$130
2	$240: (4 x $10 + $200)	$370
3	$480: (8 x $10 + $400)	$850
4	$960: (16 x $10 + $800)	$1,810
5	$1,920 (see pattern above)	$3,730
6	$3,840	$7,570
7	$7,680	$15,250

34. a. None   b. rates, conversion or dimensional analysis, distance formula   c. A. No – total trip 2.25 hours   B. 6.8 mph
d. A. You traveled downstream a distance of (6 mph + 2 mph)(.75 hours) = 6 miles. Traveling upstream at the same pace will take distance/rate = time so 6 miles/(6 mph – 2 mph) = 1.5 hours, 1.5 hours + .75 hours = 2.25 hours – you are late!   B. Your average rate upstream to make it back in 2 hours would be rate = distance/time so 6 miles/(2 hours - .75 hours) = 4.8 mph. Because the current is working against you at 2 mph, you will need to paddle at a rate of 4.8 mph + 2 mph for 6.8 mph.

35. a. None   b. conversion or dimensional analysis, geometric formulas c. Yes, you only need 3.9 ft^2 d. The total surface area of the box in inches is 2LW + 2LH + 2WH or 2(6 x 14) + 2(14 x 10) + 2(6 x 10) = 568 in^2 or (568 in^2)(1 ft^2/144 in^2) = 3.9 ft^2.

36. a. None   b. Pattern, calculation   c. A. 270,592 people
B. About 25 people   d. A.

Share Round	People in this share round	Total people view video
1	256	256
2	(256/2)(64)=8,192	8,448
3	(8,192/2)(16) = 65,536	73,984
4	(65,536/2)(4) = 131,072	205,056
5	(131,072/2)(1) = 65,536	270,952

B. (270,952 people/ 3 hours) (1 hour/60 minutes)(1 minutes/60 seconds) = 25.055 people/second or 25 people

37. a. None b. Equation, make a chart, trial and error
c. You would need a 102% on your final exam, so you cannot get an A in class if the top final exam grade possible is 100%
d. (87%)(.8 portion of class grade) + (x%)(.2 portion of class grade from final exam) = (90% hoping for)(1 total class grade), .696 + .2x = .9, .2x = .204, x = .102 or 102%.

38. a. Need average flow of average size garden hose – 5/8 inch, 100 feet long, 45 psi: water flow rate of 12 gallons per minute
b. geometric formulas, conversion or dimensional analysis
c. 20.4 hours d. Volume of pool in cubic feet = $\pi r^2 h$,
$\pi \cdot (24 \, ft \, diameter / 2)^2 \cdot (52in)(1 \, ft / 12in) = 1,960.354$ ft³. 1 ft³ = 7.481 gallons so pool holds (1,960.354 ft³)(7.481 gallons/1ft³) = 14,665.408 gallons. The pool will take (14,665.408 gallons) (1 minute/12 gallons) = 1,222.117 minutes, (1,222.117 minutes) (1 hour/60 minutes) = 20.4 hours.

39. a. Difference between low tides and high tides in VA beach are about 3 feet b. trigonometry c. The beach equipment is 2.2 feet under water. d. (see pic low tide), A grade of 15% translates to an angle of 8.851 degrees by using arctan (.15). With a hypotenuse of 18 feet, the matching vertical height is 2.670 feet by using $18 \cdot \sin 8.531°$. (see pic high tide) Notice then, that when the tide rises 3 feet vertically, that this is higher than the vertical measure of 2.67 feet at low tide, so the beach equipment must be underwater. By using $\dfrac{3 \, ft}{\sin 8.531°}$ you see that the hypotenuse is now 20.2 feet, which represents where the water now hits the beach. So, the beach equipment is 20.2 – 18 = 2.2 feet underwater!

Low tide

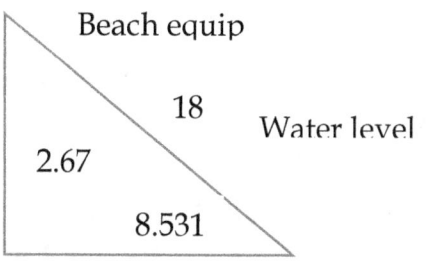

High tide

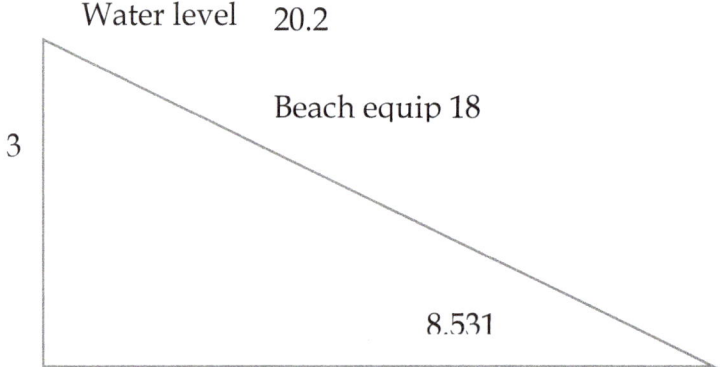

40. a. None   b. geometric formulas, conversion or dimensional analysis   c. 21 neon tetra   d. Volume of water in tank one inch from top = LWH, V=27 in x 12 in x 15 in = 4,860 in³. 1 gallon = 231 in³ so (4,860 in³) (1 gallon/231 in³) = 21.039 gallons so 21 neon tetra.

41. a. None   b. expected value, pattern or chart   c. $1 per play   d. (.1 probability of losing $5)(-$5) + (.9 probability of not losing $)($x) = $.40 expected earnings each play, -.5 + .9x = .4, .9x = .9, x = $1

42. a. You can give as much information or as little as you feel is appropriate. On the PowerBall website, how the game is played and the odds of winning the jackpot are given. All the information you need: PowerBall is played by choosing 5 numbers from 1-69 and 1 number from 1-26. The odds given on the website of winning the jackpot is 1/292,201,338. If you would like students to verify these huge odds and see where this number came from, you can use the combination formula.   b. expected value, combination formula, Fundamental Counting Principle   c. -$1.86   d. Expected earnings per play = (292,201,337/292,201,338 odds of losing $2)(-$2) + (1/292,201,338 odds of winning jackpot)($40,000,000) = -$1.86. If you would like students to compute the odds of winning the jackpot using the combination (plus a little more formula) = n!/(r!(n-r!)) where n = amount of numbers choosing from and r=amount of numbers choosing. Use this formula to find odds of matching 1ˢᵗ 5 numbers: 69!/((5!)(64!)) = 11,238,513. Then multiply 11,238,513 x 26 using the Fundamental Counting Principle = 292,201,338 so 1/292,201,338.

43. a. You need the length of each lane of a standard high school track. From inside to outside: 400m, 407.67m, 415.33m, 423m, 430.66m, 433.38m, 446m, 453.66m.   b. conversion or dimensional analysis   c. A. yes B. 4 miles   d. You ran each lane twice, except for the outside lane, which you ran only once: 2(400) + 2(407.67) + 2(415.33) + 2(423) + 2(430.66) + 2(433.38) + 2(446) + 453.66 = 6,365.74 meters. (6,365.74 meters) (1 mile/1,609.34 meters) = 3.955 or just under 4 miles.

44. a. None   b. conversion or dimensional analysis, rates
c. A. 8,096   B. 308 feet   C. A little longer than a football field
d. A. (.23)(35,200) = 8096   B. (42 miles/1 hour) (1 hour/3,600 seconds)(5,280 feet/1 mile) (5 seconds) = 308 feet   d. A football field is 300 feet, so 8 feet more

45. a. Most recent statistics: Median weekly income for high school graduates and not beyond $678 and is $1,341 for those with a master's degree. The unemployment rate for high school graduates and not beyond is 5.4% and is 2.4% for those with master's degrees.
b. arithmetic calculations   c. A. The master's degree holder earned $1,271,712 more   B. The high school graduate and not beyond is 2.25 times more likely to be unemployed
d. (65-23 years)(52 weeks/1year)($1,341/1 week) = $2,928,744 lifetime for a master's degree holder and (65-18 years)(52 weeks/1year)($678/1 week) = $1,657,032. The difference between these amounts is $1,271,712.   B. 5.4%/2.4% = 2.25 times

46. a. None   b. pattern or chart, exponential/logarithmic functions, probability   c. A. 151 pulls   B. 75.5 minutes
C. $1.32/minute   d. A. ($100 initial amount)(.97 probability)^x = $1 output amount, $.97^x$ = .01, ln $.97^x$ = ln .01, (x)(ln .97) = ln .01, x = (ln .97)/(ln .01), x = 151.191 or about 151 pulls.   B. (151 pulls) (30 seconds/1 pull) = (4,530 seconds)(1 minute/60 seconds) = 75.5 minutes   C. $100/75.5 minutes = $1.32/minute.

47. a. None   b. area model, geometry, probability   c. 5/9 or 55.6%   d. (see diagram below) Because all times, not just integer times need to be considered, an area model works well to cover all meet/not meet times. Looking at the diagram, the total area is 3,600 units^2 and the area of the two triangles that represent not meeting is (2)(.5)(40)(40) = 1,600 units^2. So the probability of not meeting is 1,600/3,600 or 4/9, and so the probability of meeting is 1-(4/9) or 5/9.

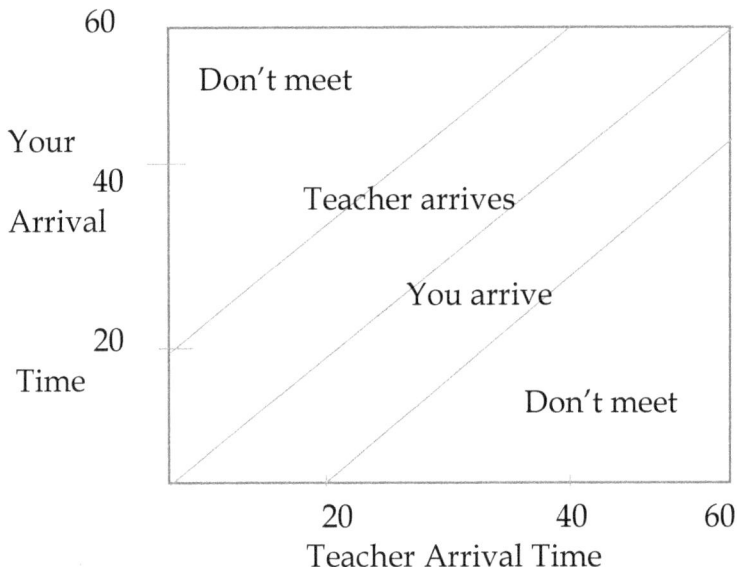

48. a. None   b. trigonometry, rates, conversion and dimensional analysis, Pythagorean Theorem, distance formula
c. A. 2,180.6 feet or .4 miles east   B. 1.7 miles   C. 14.5°
d. A. First, get time you swam to get distances using the Pythagorean Theorem and distance = rate x time: $(1.6 \text{miles})^2 + ((.5 \text{ mph})(t \text{ time}))^2 = ((2)(t \text{ time}))^2$, $2.56 + .25t^2 = 4t^2$, $2.56 = 3.75t^2$, $.683 = t^2$, t = .826 hours. This means you traveled (.5 mph) (.826 hours) = .413 miles or 2,180.6 feet. B. Your actual swimming path was (2 mph) (.826 hours) = 1.652 miles long (see diagram). C. Your actual path from north was $\sin^{-1}(.413 \text{miles}/1.652) = 14.5°$ west, so your bearing when you swim to end up exactly north would be N 14.5° E.

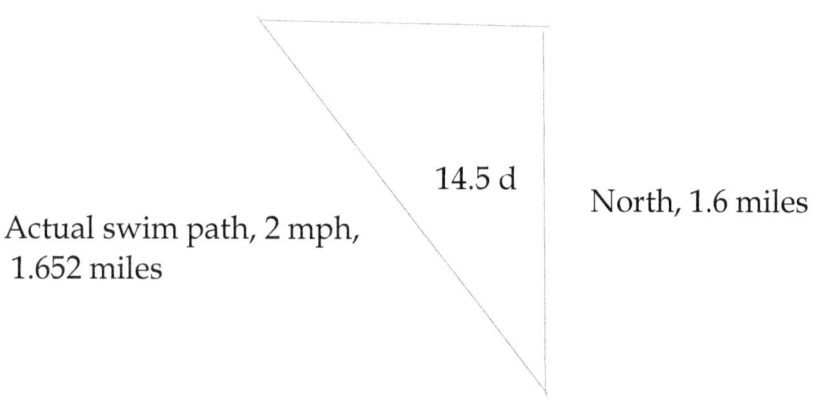

West river flow .5 mph, .413 miles

14.5 d

North, 1.6 miles

Actual swim path, 2 mph, 1.652 miles

49. a. You may use current gas prices for this problem. You will also need to know that 20 pounds of carbon dioxide emissions are produced per gallon of gas, used 13/3weeks/month
b. rates, conversion or dimensional analysis   c. A. $70.06
B. 586.3 pounds per month   C. 143.0 pounds
d. A. ((200 miles/1 week)(52 weeks))/12 months = 866.67 miles/month, (866.67 miles/1 month)(1 gallon/17 miles)($2.39/1 gallon) = $121.84/month (866.67 miles/1 month)(1 gallon/40 miles)($2.39/1 gallon) = $51.78 month, $121.84 - $51.78 = $70.06
B. (866.67miles/1 month)(1 gallon/17 miles) = 50.981 gallons, (866.67 miles/1 month)(1 gallon/40 miles) = 21.667 gallons, 50.981- 21.667 gallons = (29.314 gallons saved)(20 pounds carbon dioxide emissions/1 gallon) = 586.3 pounds   C. (21.667 gallon)(20 pounds)(100% - 67%) = 143 pounds saved

50. a. You need to know the average rate that hair grows per month: .5 inches   b. rates, conversion or dimensional analysis
c. 20 months   d. (10 inches)(1 month/.5 inches) = 20 months

51. a. The student uses their height in inches: example is 66 inches
b. use of formula   c. A. 143 pounds   B. yes, reasonable and fairly accurate   d. A. Weight = (5.5)(66inches) – 220 = 143 pounds
B. yes

52. a. None   b. proportions or rates   c. 742.583 feet
d. $\dfrac{67 inches}{12mm} = \dfrac{x}{133mm}$, (742.583 inches)(1 foot/12 inches) = 61.9 feet.

53. a. You need to know how many square feet a typical gallon of paint covers: 400 ft² and a standard wall height is 7 feet 10.5 inches.
 b. diagram, arithmetic calculations, conversion or dimensional analysis   c. 1.858 gallons so 2 gallons   d. (7.875wall height)(6 + 4 + 6 + 7 + 10 + 10 + 11 wall lengths) = 425.25 ft² total wall space.
425.25 ft² − (2)(2.667ft)(6.667ft)doors − (3ft)(6ft)window = 371.688 ft² to paint once, (371.688 ft²)(2coats) = 743.376 ft² total to paint, (743.376 ft²)(1 gallon/400 ft²) = 1.858 gallons so 2 gallons.

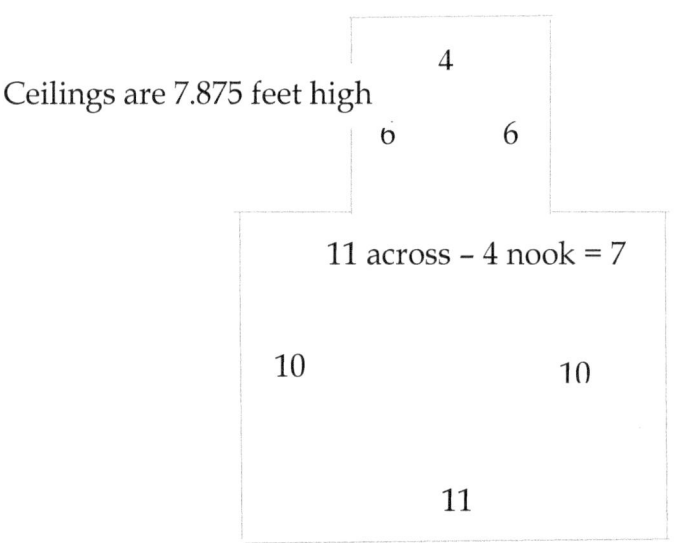

54. a. None   b. geometric formulas   c. A. 593.8 ft²   B. 256.6 ft³
d. A. Surface area = $\pi r^2 + \pi r h$, $(\pi)(7^2) + (\pi)(7)(20)$ = 593.8 ft²
B. Volume is half a cylinder: $.5\pi r^2 h = (.5)(\pi)(7^2)(20)$ = 1,539.381 ft³, (1,539.381 ft³)(1/6) = 256.6 ft³

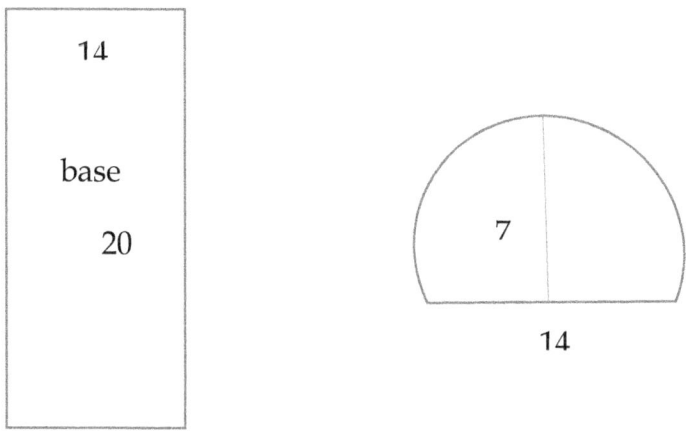

55. a. The current world population is about 7.4 billion. The current world population growth rate is about 1.1%. 24,642,757 miles² of earth is habitable. b. exponential functions, conversion or dimensional analysis   c. A. 81.4 million   B. 92,837.9 ft²   C. 203.7 year   d. A. 7.4 billion x 1.1% = 81.4 million   B. (24,642,757 mi²) (5,280² ft²/1 mi²) (1/7,400,000,000 people) = 92,837.9 ft²/person

C. $\dfrac{24{,}642{,}757 mi^2 \, x \, (5{,}280^2 \, ft^2 \,/\, 1mi^2)}{7{,}400{,}000{,}000 \, x \, (1.011)^x} = 10{,}000 \, ft^2 \,/\, person$,

$1.011^x = \dfrac{24{,}642{,}757 \, x \, (5{,}280^2)}{7{,}400{,}000{,}000 \, x \, 10{,}000}$, $1.011^x = 9.284$, ln 1.011^x = ln 9.284, x = ln 9.284/ln 1.011, x = 203.684 years

56. a. None   b. Geometric formulas and unit rates   c. A. The 16-inch is the better deal at $.07/in²   B. The medium snow cone is the better deal at $.10/in³   d. A. The 12 in pizza: total cost/1 in² = ($8.99 + $3.00)/((6 in radius)²($\pi$)) = $.11/in² and the 16 in pizza: total

cost/1 in² = ($9.99 + $4.50)/((8 in radius)²($\pi$)) = $.07/in² B. The cost of the small snow cone in $/in³ using volume of a cone formula:

$$\frac{\$1.50}{\frac{1}{3}\pi(1.5 inradius)^2 \cdot 6 inheight} = \$.11/in^3$$

and the cost of a medium snow cone in $/in³ using volume of a cone formula:

$$\frac{\$3.50}{\frac{1}{3}\pi(2 inradius)^2 \cdot 8 inheight} = \$.10/in^3$$

57. a. Students could also choose 3 of their own cities to use. Using these 3 cities, students need distances between the cities – Las Cruces, NM to Milwaukee, WI is 1,551 miles and Bismarck, ND to Milwaukee, WI is 764 miles and Bismarck, ND to Las Cruces, NM is 1,294 miles.   b. geometry formulas, drawing, program like Geometer's Sketchpad, use of a map   c. A. Wichita, KS B. 777.7 miles   d. A. Wichita, KS by drawing triangle between 3 cities and then constructing perpendicular bisectors for all three sides and locating circumcenter.
B. By measuring on map provided and using proportions or my method of using the formula for a circumcenter of a triangle where sides are length a,b,c:

$$\frac{abc}{\sqrt{(a+b+c)(b+c-a)(c+a-b)(a+b-c)}},$$

$$\frac{(1,551)(764)(1,294)}{\sqrt{(1,551+764+1,294)(764+1,294-1,551)(1,294+1,551-764)(1,551+764-1,294)}}$$

, 777.7 miles

58. a. None   b. TVM solver on graphing calculator or finance equation   c. A. 62.2 months or 5.2 years   B. $555.35
d. A. Using the formula $P = \frac{r(PV)}{1-(1+r)^{-n}}$ where P=payment ($25), r=rate per period (18.3%/12), PV=Present Value ($1,000), n=number of periods (solving for in months), $25 = \frac{(.01525)(1000)}{1-(1.01525)^{-n}}$, 15.25 = 25(1-(1.01525)$^{-n}$), 15.25 = 25-25(1.01525)$^{-n}$, .39 = 1.01525^{-n},

2.564 = 1.01525^n (reciprocal of both sides), ln 2.564 = ln 1.01525^n, n = ln 2.564/ln1.01525, x = 62.2 months or 5.2 years. Using TVM Solver where I% = 1.525, PV = 1,000, PMT = -25, FV = 0, solve for N. B. Total interest paid: (62.214months)($25) - $1,000 principal = $555.35 interest

59. a. None   b. Use of formula   c. A. 66.6%   B. 58.1%
d. A. Score (12months) = 100 − 30log(12 + 1) = 66.6%   B. Score (24months) = 100−30log(24 + 1) = 58.1%

60. a. None   b. exponential functions base e   c. 6.6 hours
d. Using the formula $y = n \cdot e^{kt}$, solve for k = rate first. $.5 = 1 \cdot e^{k \cdot 1 hour}$, .5=e^k, ln.5= lne^k, -.693 = k, so $y = n \cdot e^{-.693t}$. $1\% = 100\% \cdot e^{-.693t}$, .01 = e$^{-.693t}$, ln .01 = ln e$^{-.693t}$, ln .01/-.693 = 6.6 hours

61. a. None   b. write an equation or logic with arithmetic
c. 6 students   d. $\dfrac{\$650 + \$40s}{s} = \$50$, 50s = 650 + 40s, 10s = 650, s = 65. Currently 59 students (you and 58 others), so you need to recruit 6 students.

62. a. None   b. Use a chart/pattern or write an equation
c. A. $7 per ticket   B. $2,450   d. Using an equation where t = price per ticket, p = number of people, (t)(p) = revenue (r). First write equation for people using the point-slope formula using the points ($3,550) and ($4,500): P-500 = (-50)(t − 4) or P = (-50)t + 700. So, to maximum revenue, maximum the equation = t(-50t + 700) or r= -50t² + 700t. Using a graphing calculator or the vertex formula $(\dfrac{-b}{2a}), f(\dfrac{-b}{2a})$ where a = -50 and b = 700. So $(\dfrac{-b}{2a})$ = (-700/-100), ticket price = $7 and people = (-50)(7) + 700 = 350. B. So, total revenue would be maximized ($7/person)(350 people = $2,450.

63. a. Students would use their hometown distance from Fairbanks, Alaska. Detroit, Michigan to Fairbanks, Alaska = 3,815 miles. Earthquake tremors travel at 8 kilometers/1 second.
b. conversion or dimensional analysis   c. 12.8 minutes
d. $time = 3{,}815 miles \cdot \dfrac{1.609 km}{1 mile} \cdot \dfrac{1 \sec ond}{8 km} \cdot \dfrac{1 \min}{60 \sec} = 12.8 \min$

64. a. None   b. geometric formulas, probability   c. Only .9%

d. Probability of one run being in a 1 mile radius using the area of a circle = $\dfrac{1^2 \cdot \pi}{3 \cdot 5 \, total \, area}$ or $\dfrac{\pi}{15}$. So the probability of three runs being within a 1 mile radius is $(\dfrac{\pi}{15})^3 = .9\%$.

65. a. You need a lot of information! First 2.9% of high school men and 3.1% of high school women basketball player go on to play college ball. 1.3% of men college basketball players go on to play professionally, and 1% of women college basketball players go on to play professionally. In Michigan, there are about 687 high school basketball programs for each gender with an average of 14 members for each team.   b. rates, arithmetic computation
c. A. 298 women   B. 3 women   C. 279 men   D. 4 men   E. Yes!
d. A. (687 teams)(14 members/1 team)(3.1%) = 298 women.
B. 298 x 1% = 3 women in the whole state   C. (687 teams)(14 members/1team)(2.9%) = 279 men   D. 279 x 1.3% = 4 men in the whole state   E. Yes!

66. a. None   b. arithmetic computation, rates and percentages
c. A. Your friend by $.10 more   B. You, by 1.25%
d. A. $1.00 friend - $.90 you = $.10   B. Your friend $1/$10 = 10% increase and you $.90/$8 = 11.25%.

67. a. None   b. arithmetic computation, rates and percentages
c. A. No   B. $182.00   d. A. No   B. ($200)(1 + .30) = $260 price of bike due to demand. Then ($260)(1 - .30) = $182 price at the end of the season

68. a. The earth spins at about 1,040 miles per hour, ignore the curvature of your path/the earth   b. proportions, conversion
c. 4.3 minutes longer   d. For the sun to set in the west, the earth is spinning eastward, so you are driving west against the east spin. So, the earth is spinning east at 1,040 miles/hour – losing 75 miles/hour driving = 965 miles/hour of progress, $\dfrac{1,040 \, miles}{1 \, hour} = \dfrac{965 \, miles}{x \, hours}$, x = .928 hours, 1 hour - .928 hours = (.072 hours)(60 minutes/1 hour) = 4.3 minutes.

69. a. You can update this data each year past 2017. b. rates, percentages and arithmetic computation   c. A. $40,674.20 B. $46,674.58   d. A. $9,650 + 9,650(1.035) + 9,650(1.035)^2 + 9,650(1.035)^3 = $40,674.20   B. 9,650(1.035)^4 + 9,650(1.035)^5 + 9,650(1.035)^6 + 9,650(1.035)^7 = $46,674.58

70. a. None   b. statistics, arithmetic computation   c. A. $0 B. $0 C. $41.88 D. Opinion and great time for discussion on how data was collected, outliers, etc.   d. A. $0 is the earnings that occurs most often B. When the earnings information is listed in order, the middle or median value would be $0. C. ((130)($0) + (30)($50 interval average) + (30)($150 interval average) + $2,000))/191 people surveyed = $41.88.

71. a. None   b. arithmetic computation and rates   c. A. $1,352 B. 150.2 hours or 7.5 weeks   d. A. ($4latte)(4 times per week)(52 weeks/1year) + ($20total movie cost)(26times per year) = $1,352. B. ($1,352)(1 hour/$9) = 150.2 hours or (150.2 hours)(1 week/20 hours) = 7.5 weeks.

72. a. The number of students at your high school: example 1,300. The average price of a pack of cigarettes in your state: example Michigan $6.64.   b. rates, conversion   c. A. 77 students B. 26 students C. $2,423.60 per year   d. A. (1,300 students)(5.9%) = 77 students B. (77 students)(1/3) = 26 students C. (365 packs per year)($6.64/1 pack) = $2,423.60 per year

73. a. For 2016, Social Security rate is 6.2%, Medicare rate is 1.45%, state tax rate (for Michigan) is 4.25% and the Federal tax rate for the lowest tax bracket is 10.6%.   b. percentages and arithmetic computation   c. A. $279.00 B. 9 hours   d. A. ($360)(.062 + .0145 + .0425 + .106) = $81 in taxes, so $360 - $81 = $279   B. ($81)(1 hour/$9) = 9 hours

74. a. Current number of Las Vegas residents is 633,000. b. conversion, scientific notation   c. A. 5.9 x 10^{11} gallons per year B. 19.9 years C. 16.6 years D. 55.4 million gallons d. A. (247 miles^2)(5,280^2 ft^2/1mile^2)(1,296 – 1,079 change in feet) = 1.494 x 10^{12} ft^3, (1.494 x10^{12} ft^3)(7.48052 gallons/1 ft^3) = (1.118 x 10^{13})/(2017-1998 years) = 5.883 x 10^{11} gallons per year. B. Loss of feet/year = (1,079 – 1,296 feet)/(2017 – 1998) = -11.421 feet in height/year, (227 feet remaining)(1 year/11.421 feet) = 19.9 years. C. (190 more feet)(1 year/11.421 feet) = 16.6 years D. (125 gallons)(633,000 people)(70%) = 55.4 million gallons

75. a. None   b. probability, percentages   c. A. 4 students
B. 4 students C. 2 students   D. 6 students E. 20 students
F. 7 students G. 100%   d. A-F: Multiply 30 students by each percentage

76. a. None   b. arithmetic computation   c. 1st Place: Darlene, 2nd Place: Carl, 3rd Place: You! Congratulations!   d. Points for You: 8(.5) + 3(.3) + 7(.2) = 6.3, Butch: 5(.5) + 9(.3) + 4(.2) = 6, Carl: 9(.5) + 6(.3) + 3(.2) = 6.9, Darlene: 8(.5) + 5(.3) + 9(.2) = 7.3, Evan: 4(.5) + 8(.3) + 7(.2) = 5.8

77. a. There are approximately 9 games per season and about 57 practices and approximately 1,088,158 high school football players in the US currently per season   b. rates, percentages, conversion
c. A. 5.1% B. 1.7% C. 54,496   d. A. (66 engagements)(76.8/100,000 concussion rate) = 5.1% B. (5.1%)(33%) = 1.7%
C. (1,088,158 football players)(5.1%) = 55,496

78. a. Caffeine in mgs in Monster Energy drink: 160 mg, 12-oz Coke: 34 mg, Starbuck's Grande brewed coffee: 330 mg   b. arithmetic calculations   c. A. .625 of a Monster B. 2.9 Cokes C. .3 of a Starbucks   d. A. (100mg)(1 Monster/160 mg) = .625 of a Monster
B. (100 mg)(1 Coke/34 mg) = 2.9 cans of Coke C. (100 mg)(1 Grande/330 mg) = .3 of a Grande

79. a. A water tower provides .43 psi per foot in height from ground level, and the average household uses 90 gallons per day.
b. arithmetic computation   c. A. 154 feet B. 1.1 million gallons
d. A. ((75 psi)(1 foot/.43i psi)) − 20 ft hill = 154 feet B. (13,000 households)(90 gallons/1 household each day) = 1,117,000 gallons

80. a. None   b. formulas, graphing calculator, may use calculus if appropriate, use of diagram   c. A. 1,879.7 feet or .356 miles
B. 4.7 minutes   d.   A.-B.   Write an equation with respect to time first: total time = (distance run on beach)/(12 mph) + (distance swim to victim)/(6 mph), time = $\dfrac{x}{12} + \dfrac{\sqrt{(.5-x)^2 + (.25)^2}}{6}$. Use graphing calculator to graph function and find where function is at a minimum remembering x is less than .5 miles and time should be a small part of an hour: x=.356 miles and y = .078 hours or x = 1,879.7 feet and y = 4.7 minutes.

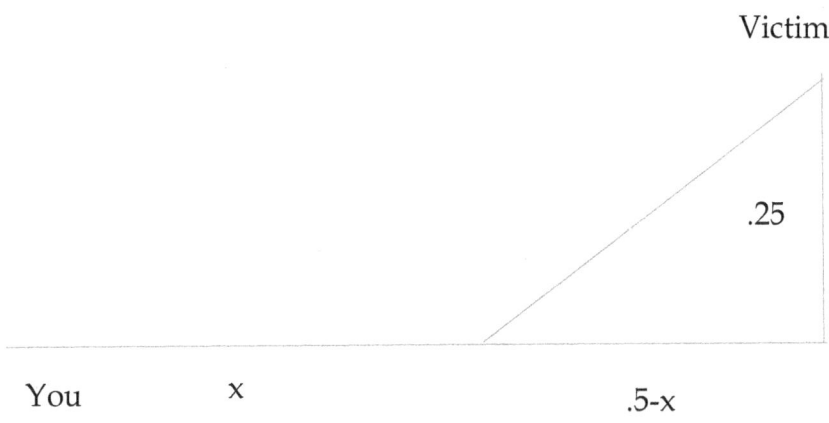

81. a. None   b. drawings, chart or list   c. You would purchase less feet of wood and waste less by buying 29 2 x 4 x 8 planks of wood   d. Each of four walls with window and door are easy to draw and visualize, so are not provided here. Use wood efficiently and remember each foot in height of wall requires 3 boards (width of 4 inches each). Efficient lists for 2 x 4 x 8 and 2 x 4 x 10 are below:

2 x 4 x 8 – total purchased 30 planks so 240 feet
*For 2 smaller walls (4 ft long): (2 walls)(3 boards)(5 feet)/2 parts per plank = 15 boards
*For tops and bottoms (full lengths) of longest walls (5 ft long): (2 tops)(3 boards)(1 foot) + (3 boards)(2 feet) for under window = 12 boards used with 12 3 ft pieces left over
*Around window: (2 sides)(3 boards)(2 feet) = 12 1.5 pieces needed so used 6 3 ft pieces already available
*Around door: (2sides)(3 boards)(4 feet) = 24 1.5 pieces needed so used 6 3 ft pieces already available and need 12 1.5 pieces from 3 more boards.

2 x 4 x 10 – total purchased 25 planks so 250 feet
*For tops and bottoms (full lengths) of larger walls: (3 boards)(4 feet)/2 parts per plank = 6 boards used
*For smaller walls: (2 walls)(3 boards)(5 feet)/2 parts per boards = 15 boards used with 15 2 ft pieces left over
*Around window: (2 sides)(3 boards)(2 feet) = 12 1.5 pieces needed so used 12 2 ft pieces already available
*Around door: (2 sides)(3 boards)(4 feet) = 24 1.5 pieces needed. Used 3 2ft pieces already available then cut 4 boards for 6 1.5 pieces each board.

82. a. None   b. exponential, pattern or chart   c. A. 10 weeks ago B. 5 people   d. A.-B. First, the number of new cases tripled from one week to the next, so this is an exponential function where the base is 3. 295,245 can be divided in succession by 3 ten times with an integer remainder of 5. So the researcher had 295,245 cases in Week 10 and the number of people originally affected was 5.

83. a. None   b. proportions with dimensions   c. A. The larger pizza is the better deal B. The larger cube is the better deal.
d. A. The 1-foot square pizza has 1 ft² to eat for 1 unit of cost (exact cost not given). The 2-foot square pizza has 4 ft² of area to each (quadruple, not double) for 2 units of cost. So, the 1-foot square pizza is 1 unit $/1 ft² =1 units $/ft² and the 2-foot square pizza is 2units $/4 ft² = .5 units $/ft². B. The small cube (with sides of x long) and at a cost of $10 is $10/x³ in price. The larger cube (with sides 2x long) and at a cost of $20 is $20/(2x)³ or $20/8x³ or $2.5/x³. So, the larger cube is the better value!

84. a. No, as a side note, it is cool to have students look up what makes a graph traversable and create their own traversable/nontraversable graphs!   b. graph theory, logic, drawing
c.-d. A. not efficient or traversable B. example (start walking at A or B):

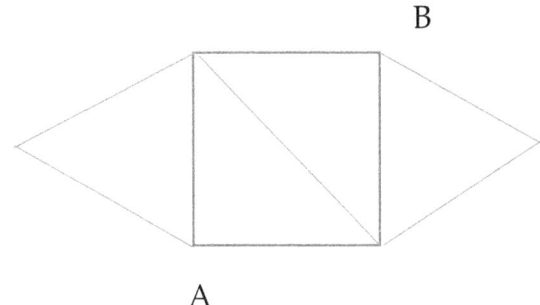

85. a. None   b. waiting line models, discrete math, scheduling
c. 14 servers   d. Using the formula Line Wait Time =

$$10 = \frac{1}{(servicerate)(servers) - (arrivalrate)} \cdot \frac{arrivalrate}{(servicereate)(servers)}$$

where service rate is the unit of time per customer per server, arrival rate is the number of customers arriving per unit time and servers = x. First service rate is 1 customer/7 minutes or .143 customers per minute. Next, arrival rate is 325 customers/3 hours or 1.806 customers per minute. line wait time = 10 minutes, and you are solving for x=servers. So, $10 = \frac{1}{(.143)(x) - (1.806)} \cdot \frac{1.806}{(.143)(x)}$, taking the

reciprocal and distributing: $.1 = \frac{.020x^2 - .258x}{1.806}$, $.1806 =$

.020x² - .258x, .020x² - .258x - .1806 = 0. Using the Quadratic Formula: x = 13.6 servers or 14 servers

86. a. The minimum distance to an outfield fence on a high school baseball field is 320 feet.   b. physics formulas, conversions   c. A. Yes B. by 4.5 feet d. A. Using the parametric equations in feet per second y height = $-16t^2 + V_0 \cdot \sin\theta \cdot t + h$ and x horizontal distance = $V_0 \cdot \cos\theta \cdot t$: First convert V_0 = 98 miles/hour to 143.733 feet/second. Also note that sin 45 degrees and cos 45 degrees is $\frac{\sqrt{2}}{2}$

and h=5 feet. Now, because air resistance causes the ball to only get ½ the horizontal distance, by transformation, the horizontal distance equation is now 2x = $V_0 \cdot \cos\theta \cdot t$. First find the time x would be 320 feet to the fence: (2)(320) = (143.733)($\frac{\sqrt{2}}{2}$)(t), 640 = 101.635t, t = 6.297 seconds. So, the ball would be at the fence line at 6.297 seconds. Input this time into y height =

$-16(6.297)^2 + (143.733)\frac{\sqrt{2}}{2} \cdot (6.297) + 5$, y = 10.461 feet. B. The ball clears the fence by 10.461 – 6 feet = 4.5 feet.

87. a. None   b. trial and error with logic   c. A. examples: 341 + 586 = 927 or 143 + 586 = 729 or more!   B. No
d. Notice, if you are carrying a 1 into the hundreds place, your two numbers left should be two values apart. Try consistent strategies – middle numbers or large/small together. Great opportunity to talk strategy with students.

88. a. None   b. conversion   c. 500,000 kg or 1.1 million pounds
d. $(1 \text{ km}^3)(1{,}000^3 \text{ m}^3/1 \text{ km}^3)(.5 \text{ grams}/1 \text{ m}^3 \text{ water}) = 500$ million grams water. (500 million grams)(1 kg/1,000 g) = 500,000 kgs. (500 million g)(1 pound/453.592 g) = 1.1 million pounds

89. a. The average height of a woman is 5 ft 4 in or 64 in.
b. geometric formulas   c. 33.5 ft only!   d. Using circumference = $2\pi r$: $2\pi(r + 64)$ circumference of head - $2\pi r$ circumference of feet = $2\pi \cdot 64 = 402.124$ inches, (402.124 inches)(1 foot/12 inches) = 33.5 feet

90. a. None   b. arithmetic computation   c. 126° strange!
d. 26(position of minute hand) − 5(position of hour hand) = 21 mark difference. Each mark on clock is $\dfrac{360°}{60}$ = 6°, so (21 marks) (6°/1 mark) = 126°.

91. a. None   b. trial and error with logic   c. A. examples 1 + 2 + 3 − 4 + 5 + 6 + 78 + 9 + 0 = 100 or 1 − 2 + 3 − 4 + 5 + 6 − 7 + 8 +90 = 100
B. Yes   d. One strategy would be to form a two-digit number close to 100 and add/subtract smaller numbers to reach 100 exactly

92. a. None   b. pattern with starting small   c. 1
d. Begin with exponent one, continue raising exponent by 1 higher value while recording units place of solution. Spot the units place pattern and use it to determine the unit place of 3^{1000}.

93. a. None   b. rates   c. 26.7 mph (not 30 mph!)
d. Average speed = total distance/total time, total distance was (2 trips)(40 miles/trip) = 80 miles, total time for the trip is 1 hour on way to grandma's house + 2 hours on way home = 3 hours total. So, the average speed is 80 miles/3 hours = 26.7 mph.

94. a. None   b. pattern, chart or list   c. A score of 43 is not possible on this quiz   d. The highest score is, of course, 50 points. Following is a chart that calculates next highest scores that shows that 43 is not a possible score:

Correct (5 pts)	Blank (1 pt)	Incorrect (0 pts)	Total score
9	1	0	46
9	0	1	45
8	2	0	42
8	1	1	41

95. a. None   b. trial and error with logic   c. A. $3 \times 54 = 162$ B. Yes   d. There are not as many multiplications to check as you would think. The first number cannot be 1,2 (won't get a 3-digit output) or 5 (would give a units place of 5 or 0 in solution). The tens place of the 2 digit number cannot be a 1. The units place of the 2-digit number cannot be 1 or 5. The first digit of the solution cannot be 4,5,6 because no answer can possibly be higher than 300. The units place of the solution cannot be 1 or 5. Then, remember any multiplication has to produce a 3-digit number and no repeating numbers that you used to multiply with already.

96. a. None   b. pattern or chart, equations, use of graphing calculator   c. A. $250,000 B. $10.7 million C. Days 24-25 d. Through use of equations and a graphing calculator: A. Reward = 100,000 + 5,000(days), so when days = 30, the reward is 100,000 + (5,000)(30) = $250,000 B. Reward = $.01(2)^{days}$, so when days = 30, the reward is $.01(2)^{30}$ = $10,737,418.24 C. Using graphing calculator to graph both equations, Option 2 became better than Option 1 between days 24-25.

97. a. None b. Make a list, probability theory   c. A. No B. 1/8 C. 1/8 D. ¼   d. By a using a list of all possible outcomes: EEE,EEO,EOE,OEE,EOO,OEO,OOE,OOO, A,B,C. Notice 1/8 possibilities is all evens, and 1/8 possibilities is all odds. D. Notice 2/8 possibilities or ¼ has all dice the same. By using probability: A,B,C. $(1/2)^3$ = 1/8 for all evens and same for all odds. D. $2 \cdot (1/2)^3$ = ¼ or 1/8 + 1/8 = ¼.

98. a. Average life span of US woman is 81, for US men 76. Average heart rate is 80 beats per minute. Average weight of food eaten per day is 4 pounds. Average weight of trash produced is 4.3 pounds per day. Average amount of sleep is 8.45 hours per night.

b. rates, conversion   c. A. Women: 3.4 billion beats Men: 3.2 billion beats B. Women: 2.6 billion secs and Men: 2.4 billion secs. C. Women: 118,260 pounds and Men: 110,960 pounds D. Women: 127,129.5 pounds and Men: 119,282 pounds E. Women: 249,824.25 hours or 28.5 years and Men: 234,403 hours or 26.8 years F. Go for it!
d. A. Women: (81 years)(365 days/year)(24 hours/day)(60 minutes/hour)(80 beats/min) = 3,405,888,000 beats and Men: same math but using 76 years instead of 81 years. B. (81 years)(365 days/year)(24 hours/day)(60 minutes/hour)(60 seconds/min) = 2,554,416,000 seconds and Men: use 76 years C. (81 years)(365 days/year)(4 pounds/day) = 118,260 pounds and Men: use 76 years D. (81 years)(365 days/year)(4.3 pounds/day) = 127,129.5 and Men: use 76 years E. (81 years)(365 nights/year)(8.45 hours/night) = 249,824.5 and Men: use 76 years F. Go for it!

99. a. None   b. pattern, summation formula   c. 500,500
d. By pattern, notice that if you start at the smallest numbers and largest numbers that 1+999, 2+998, 3+997, 4+996, etc. all add to 1,000 – this happens 499 times. So, (499)(1,000) + 1,000 (largest number) + 500 you never used in pairs = 500,500. Or, by summation formula: $\sum_{1}^{n} n = n(n+1)/2$, so 1000(1001)/2 = 500,500.

100. a. Population density of NYC: 27,016 people per square mile. City on planet with highest population density: Dhaka, Bangladesh with 112,700 people per square mile. Information on your city example Detroit, MI: 5,142 people/square mile.   b. conversion
c. A. 1,031.9 ft^2/person B. 247.4 ft^2/person C. 4.2 times more dense D. Measure! E. 5,421.7 ft^2/person F. 5.3 times less dense
d. A. (1 mi^2/2,7016 people)(5,280^2 ft^2/1 mile^2) = ,1031.9 ft^2/person B. (1 mi^2/ 112,700 people)(5,280^2 ft^2/1 mile^2) = 247.4 ft^2/person C. 112,700/27,106 = 4.2 E. (1 mi^2/5,142 people)(5,280^2 ft^2/1 mile^2) = 5,421.7 ft^2/person F. 27,016/5,142 = 5.3

101. a. Your total hours in school over 4 years – usually set by each state. Example: Michigan 180 days per year, 7 hours per day.
b. rates, conversion   c. A. $53.97/hour B. Be honest!
d. ($272,000)/((4 years)(180 days)(7 hours) = $53.97 per hour

102. a. None   b. trial and error with logic c. A.   3   B. Yes C. Same grouping of values

d. B. Of course! Due to symmetry

```
    3
   4 5
  2 6 1
```

103. a. None   b. rates, conversions   c. No – they could not sell 3.5 million hot dogs   d. Using reasonable assumptions: 1 employee does all the managing, cleaning, customer care at the least, the other 5 employees only cook hotdogs. (5 employees) (40 hours/week)(52 weeks/year)(60 minutes/1 hour)(60 seconds/hour) allows 37,440,000 seconds of employee time to make hotdogs. This allows (37,440,000 seconds)/(3,500,000 hotdogs) = 10.7 seconds per hotdog. This is not enough time to fully cook and prepare a hotdog! Note-If other reasonable assumptions are made/allowed, the answer may become plausible.

104. a. None   b. prime factorization, trial and error   c. (16)(625)   d. Using prime factorization: 10,000 = **(2)**(5,000) and 5,000 = **(2)**(2,500) and 2,500 = **(2)**(1,250) and 1,250 = **(2)**(625) and 625 = **(5)**(125) and 125 = **(5)(5)(5)**. So, 10,000 = $(2^4)(5^4)$ or (16)(625).

105. a. None   b. trial and error with logic   c. F=5, G=4, H=6, J=0, K=9   d. First notice in the units place that H – H must = 0 for J. Second, in the tens place, since J=0, you must carry from H in the 100s place and 10 – 5 = 5, making F=5. Third, in the 10,000s place, G must be one value less than F since J=0. Other clues, H must be one more than F due to 100s place, and G is less than K in the 1000s place.

106. a. None b. trial and error with logic   c. and d. First fill 5 gallon bucket with water and pour into 8 gallon bucket. Refill the 5 gallon bucket and use this water to fill 8 gallon bucket to the top. 2 gallons are now remaining in the 5 gallon bucket. Add mix to make lemonade.

107. a. None   b. trial and error with logic, use bingo chips   c.-d. Force person to choose from an amount of chips remaining that is exactly one more than a multiple of 5. Example: You take 4 leaving 31 (1 more than multiple of 5). Friend takes 2 leaving 29. You take 3 leaving 26. Friend takes 4 leaving 22. You take 1 leaving 21. Friend takes 3 leaving 18. You take 2 leaving 16. Friend takes 1 leaving 15. You take 4 leaving 11. Friend takes 4 leaving 7. You take 1 leaving 6. Friend takes 1 leaving 5 (because if they take more, you win). You take 4 leaving 1 and you win!

108. a. None   b. trial and error with logic   c. A. $(9/9) + (9)(9) + 9 + 9 = 100$ B. Yes!   d. A. Used multiplication to get close to 100 and then work with nine from there. B. Just the fact that $(9/9) = 1$, allows you to make 100 simply by adding $(9/9)$ 100 times!

109. a. None   b. pattern, inverse operations   c.-d. A. Your phone number! B. The math you are asked to do has 3 purposes – one is to enter your phone number, the second is to put the digits in correct positions, and the third are inverse operations meant to create magic. Putting in the first three digits of your phone number, and then multiplying by 80, adding 1 and multiplying by 250 doubles your digits and places 4 zeros to the right of your three digits to allow space for the last 4 digits. The adding 1 and then multiplying by 250 is undone later by subtracting 250. The dividing by 2 undoes the doubling of the first three digits and the adding of the last 4 digits twice.

110. a. None   b. rates, conversion, diagram   c. A. 550 feet   B. 3,300 feet   d. A. The train is "5 seconds" long, because that is how long it takes to enter the tunnel from beginning to end. We just have to convert 5 seconds to feet: (5 seconds)(75 miles/ 1 hour)(5,280 feet/1 miles)(1 hour/3,600 seconds) = 550 feet. B. If it takes 40 seconds for the train to completely pass through the tunnel, then the length of the tunnel in seconds is actually (40 seconds – 5 seconds for train to fully enter tunnel – 5 seconds for train to totally leave tunnel) = 30 seconds. To convert 30 seconds to feet: (30 seconds)(75 miles/1 hour)(5,280 feet/1 miles)(1 hour/3,600 seconds) = 3,300 feet.

111. a. None   b. trial and error with logic   c.-d. 7

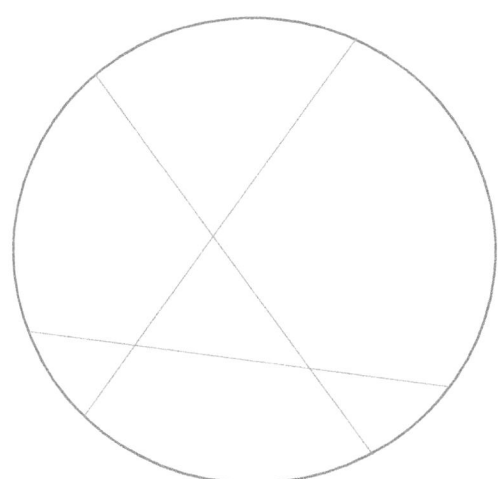

112. a. It took 20 years to build the pyramid with 100,000 slaves, the pyramid is 482 feet in height, with a square base of 756 foot long sides. More obscure data is provided.   b. conversion and geometric formulas   c. A. 513,553 blocks B. 4.8 million tons C. 238,801.9 tons per year D. 95,520.8 pounds

d. A. Number of blocks = Volume of pyramid ($\frac{1}{3} \cdot h \cdot s^2$) in feet³ of blocks in feet. The volume of the pyramid = (1/3)(482 ft)(756 ft)² = 91,826,784 ft³. The volume of each block where 1 meter = 3.281 feet: (1.5 m)(1.5 m)(2.25 m)((3.281 ft)³/1 m³) = 178.807 ft³. So number of blocks = 91,826,784 ft³/178.807ft³ = 513,553 blocks.
B. (513,553 blocks)(9.3 tons/1 block) = 4,776,038.4 tons
C. 4,776,038.4 tons/20 years = 238,801.9 tons per year
D. (4,776,038.4 tons)(2,000 pounds/1 ton)/(100,000 people) = 95,520.8 pounds per person.

113. a. The current world population is 7.5 billion people and the average life expectancy of a human being is 71.5 years.
b. probability theory and conversion   c. A. 8.1×10^{67} arrangements B. 1.7×10^{19} shuffles C. Approximately 0 (2×10^{-47}%)
d. A. Using permutation or fundamental counting principle: 52! or 8.1×10^{67} b. (7.5 billion people)(71.5 years/1 person)(365 days/1 year)(24 hours/1 day)(60 minutes/1 hour)(60 seconds/1 minute) = 1.7×10^{19} C. $1.7 \times 10^{19}/8.1 \times 10^{67}$ = 2×10^{-49} or 2×10^{-47}% or nearly zero!

114. a. LeBron James earns $33.29 million for the 2016-2017 season and there are 82 games at 48 minutes each for 6 months.
b. conversion   c. A. $8,457.83 per minute B. $501.84 per minute
d. A. ($33.29 million)/(48 minutes x 82 games) = $8,457.83 per minute. B. Total minutes of work is (26 weeks)(40 hours/1 week)((60 minutes/1 hour) + (82 games)(48 minutes/1 game) = 66,336 minutes per season. $33.29 million/66,336 minutes = $501.84/minute.

115. a. There are about 12.46 million households in the U.S. in 2016.
 b. conversion   c. 38,938 wind turbines   d. (12,460,000 households)(10,000 KWH/1 household)(1 wind turbine/3,200,000 KWH) = 38,938 wind turbines

116. a. A white oak tree has a growth factor of 5.   b. conversion
c. A. 100.3 years B. 84 feet   d. A. ((63 in/π)(5 growth factor) = 100.3 years B. (63 in)(16 height factor) = 1008 in, (1008 in)(1 ft/12 in) = 84 feet

117. a. Students use their height, example 5.5 feet, average two-story house is 26 feet high   b. conversion, proportion   c. A. 93.8 times B. 515.9 feet C. 19.8 times   d. A. (563mm/6mm) = 93.9 B. (5.5 ft)(93.9) = 515.9 feet C. (515.9 ft)(1 house/26 ft) = 19.8 times higher than a two-story house!

118. a. The height of the tallest falls in the world -- Angel Falls is 3,212 feet and Niagara Falls are 167 feet   b. conversion and physics formulas   c. A. 19.2 times higher B. 14.2 seconds   d. A. 3,212 feet/167 feet = 19.2 times higher B. Use the formula h(t) = -16t² + V₀t + h to find the time of the fall where h is height in feet and V₀ is the initial velocity in feet/second: 0 height = -16t² + (0 velocity)(t) + 3,212 so -3,212 = -16t², 200.75 = t², t = 14.2 seconds

119. a. The current national debt per citizen is $61,346 b. exponential functions   c. 82.8 years   d. 10 = 61,346(100% - 10%)ˣ, .000163 = (.9)ˣ, Ln .000163 = (x)(ln .9), x = 82.8 years

120. a. None   b. combinations, or Binomial Theorem or graphing calculator program   c. A. No   B. 11.2%   d. A. No B. Using the combination formula C = $\dfrac{n!}{r!(n-r)!}$, where n is the total you are choosing from and r is the amount you are choosing, C = $\dfrac{50!}{(25!)(25!)}$ = 1.261 x 10¹⁴, there are 1.261 x 10¹⁴ ways to toss a coin 50 times and obtain an even number of heads and tails. Applying the binomial theorem for the term (₅₀C₂₅)(.5 probability of heads)²⁵(.5 probability of tails)²⁵ = 11.2%

121. a. None   b. geometric formulas   c. A. 1.320 times B. 1.153 times   d. A. Using the formula for the volume of a cone V = $\dfrac{\pi}{3}r^2 h$ and where radius = (height/tan 38 degrees) for the dirt pile and radius = (height/tan 27 degrees) for the same volume wheat pile. V dirt = $\dfrac{\pi}{3}(\dfrac{h}{\tan 38°})^2 \cdot h$ and V wheat = $\dfrac{\pi}{3}(\dfrac{h}{\tan 27°})^2 \cdot h$, so

V dirt = 1.7557h³ and V wheat = 4.0336h³, 1.7557h³ dirt= 4.0336h³ wheat, so the dirt pile is always $\dfrac{\sqrt[3]{4.0336}}{\sqrt[3]{1.7557}}$ times taller than the wheat pile = 1.320 B. Using the formula for the volume of a cone V = $\dfrac{\pi}{3} r^2 h$ and where the height = $r \cdot \tan 38°$ for the dirt pile and height = $r \cdot \tan 27°$ for the same volume of wheat pile. V wheat = $\dfrac{\pi}{3}(r)^2 (\tan 27° r)$ or $\dfrac{\pi}{3}(r)^3 (\tan 27°)$ so radius wheat = $\sqrt[3]{\dfrac{3V}{\pi \tan 27°}}$.

V dirt = $\dfrac{\pi}{3}(r)^2 (\tan 38° r)$ or $\dfrac{\pi}{3}(r)^3 (\tan 38°)$ so radius dirt = $\sqrt[3]{\dfrac{3V}{\pi \tan 38°}}$. So the wheat pile radius compared to the dirt pile radius Is ($\sqrt[3]{\dfrac{3V}{\pi \tan 27°}}$)(1/$\sqrt[3]{\dfrac{3V}{\pi \tan 38°}}$) or 1.153 times.

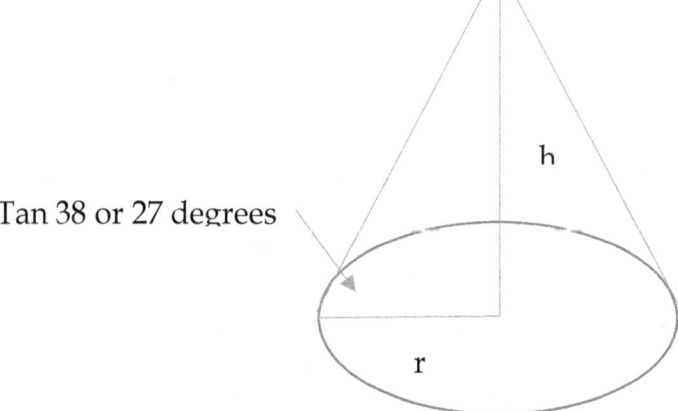

Tan 38 or 27 degrees

122. a. The formula for Fahrenheit temperature from counting cricket chirps = (# chirps/14 seconds) + 40.   b. simple arithmetic, proportions   c. 70 degrees Fahrenheit.   d. First, number of chirps in 14 seconds: $\dfrac{128 chirps}{60 \sec onds} = \dfrac{x}{14 \sec onds}$, x = 30 chirps. So 30 chirps + 40 = 70 degrees Fahrenheit.

123. a. None   b. physics/parametric formulas and conversion
c. A. 11.456 seconds B. 2,520 feet   d. A. Using the
physics/parametric formulas h(t) vertical distance =
$-16t^2 + V_0 \cdot \sin\theta \cdot t + h$, where V_0 = initial velocity in feet per second
and h = initial height in feet and t = time in seconds: First note that
V_0 in feet per second = (150 miles/1 hour)(5,280 feet/1 mile)
(1 hour/3,600 seconds) = 220 feet per second and that h(t) = 0 when
the crate hits the ground. Next, $0 = -16t^2 + 220 \cdot \sin 0° \cdot t + 2,100$, -
2,100 = -16t², t² = 131.25, t = 11.456 seconds.
B. $x(t) = 220 \cdot \cos 0° \cdot 11.456$, x(t) = 2,520 feet

124. a. The G-forces a typical human being can endure for a short
period of time is 5Gs vertically and 20Gs horizontally. The typical
mechanical watch can withstand 5,000Gs.   b. simple arithmetic
c. A. 250 times B. Why do we need watches to survive more shock
than us when it is on our wrist?   d. A. 5,000Gs/20Gs = 250 times
B. But why?

125. a. Students will need to find information they believe is
pertinent to them estimating the answer to each question.
b. conversion, simple arithmetic   c. A. A Fermi question uses
approximations and conversion to estimate as accurately as
possible an answer to the question posed. The idea is that some
approximations will be over and some under, producing perhaps a
fairly reasonable solution. B. 2.3 million pieces of paper C. 69,719
tacos D. 12,311 dentists   d. B. (Number of pieces of paper an
average student uses each day)(number of students)(number of
days in school year): (9 papers/1 day/1 student)(1,450 students/
1 school)(180 days/1 school year) = 2,349,000 pieces of paper
C. (Number of tacos eaten U.S. last year)(1/U.S. population)(number
of residents in your city)(1 year/12 months): (4.5 billion tacos/
1 year)(1/323.1 million U.S. population)(60,070 city population)
(1 year/12 months) = 69,762 tacos   D. (average number of dentist
appointments per year per person in U.S.)(state population)
(1/number of appointments dentists do on average per year):
(2 appts per year/1 person)(9.91 million people/1 state)(1
dentist/1,610 appointments) = 12,311 dentists where 1,610
appointments estimated by dentists working 46 weeks per year with
35 hours per week on appointments with 1 hour per appointment.

www.ingramcontent.com/pod-product-compliance
Lightning Source LLC
Chambersburg PA
CBHW050017230526
45470CB00003B/1012